Edward Day McNicoll, David Hudson McNicoll

Handbook for Southport

Medical and General, With Copious Notices of the Natural History of the District

Edward Day McNicoll, David Hudson McNicoll

Handbook for Southport
Medical and General, With Copious Notices of the Natural History of the District

ISBN/EAN: 9783337025595

Printed in Europe, USA, Canada, Australia, Japan

Cover: Foto ©Andreas Hilbeck / pixelio.de

More available books at **www.hansebooks.com**

HANDBOOK FOR SOUTHPORT,

MEDICAL AND GENERAL,

WITH

COPIOUS NOTICES OF THE NATURAL HISTORY OF THE DISTRICT

EDITED BY EDWARD DAY McNICOLL,

HONORARY SURGEON TO THE SOUTHPORT INFIRMARY; FORMERLY
RESIDENT MEDICAL OFFICER TO THE SOUTHPORT
CONVALESCENT HOSPITAL.

THIRD EDITION.

SOUTHPORT:
ROBERT JOHNSON AND CO, LIMITED, "VISITER" OFFICE, LORD STREET.
1883.

PREFACE.

THE first edition of this little book was compiled by my father in 1859. The sale was rapid, and in 1861 another edition was called for and soon exhausted.

That no work dealing with the same subject has equalled my father's in scope and completeness has been freely admitted. To obtain a copy—although frequent attempts to do so have been made by the public—has for many years been impossible. My father's long illness, followed by his lamented death in 1868, doubtless helped to prevent a renewed issue.

The approaching visit of the British Association to Southport is an event of importance so great, and is so likely to attract general attention to the town and its history, as well as to promote local interest in all pertaining to it, that a more fitting opportunity could hardly be found for now publishing a new edition. Although the form of the work remains the same, it is hoped that in substance it may lay some claim to be considered much improved. The remarkable, almost unexampled, progress and development of Southport, during the last twenty years, has called for a total re-casting and re-writing of much that the former editions contained—the portions, in particular, which are descriptive of the public buildings and institutions. Many of these have been founded or enlarged within the period named.

The preface to my father's original work contained a paragraph, the re-printing of which, *verbatim*, will indicate why even increased space is given to the Natural History.

"I need not apologise for the Natural History occupying so large a portion of the following pages. The motive has been to encourage useful and agreeable mental occupation on the part of visitors whose stay in Southport is more or less prolonged, and whose minds would, in the absence of some external object of thought, turn and prey upon themselves. This continual contemplation of their own condition—the result of the depression dependent upon disease, and absence from the ordinary engagements of life—needs to be carefully guarded against, and I have not thought it out of place in a work partly medical to suggest a substitute."

With a view to promoting my father's excellent object in the best manner possible, I have sought the assistance, whilst preparing this new edition, of some of the gentlemen to whom he was himself indebted for help. In this connection it gives me pleasure to name Mr. Charles H. Brown, to whom my thanks are due for undertaking the chapters upon the Shells, the Zoophytes, and the Foraminifera; and Mr. Leo Grindon, who has dealt with the Botany, and supervised the Natural History in general.

Mr. Baxendell, the well-known meteorologist, has kindly supplied me with some Tables, which present, in the clearest manner, the results of the records daily made at the Observatory in Hesketh Park, extending over a period of eleven years. These Tables have not before been published in the present form, and are a distinct and valuable feature of the volume.

To Dr. Vernon, our valued Sanitary Medical Officer, my thanks are also due for placing at my disposal information which he had collected for independent purposes.

E. D. McN.

SOUTHPORT,

AUGUST, 1883.

CONTENTS.

	PAGE
Chapter I.—Origin and Growth of Southport	1
II.—Southport as a Resort for Invalids.—Geology of the District—General Remarks on Climate—Local Climate of Southport	21
III.—Effects of the Climate upon Disease.—General Claims of Southport as a Sanatorium—Suggestions for Invalids	35
IV.—On Sea-Bathing	54
V.—Natural History of Southport and its Environs	64
The Flora	65
VI.—The Southport Birds	89
VII.—Arachnida and Crustacea of Southport	103
VIII.—Mollusca of Southport	114
IX.—Zoophytes of Southport	140
X.—Foraminifera of Southport	158
Meteorological Tables	166
Population—Table of Increase	172
Comparative Death-Rate Table	173

CHAPTER I.

―― Through days and weeks
Of hope, that grew by stealth,
How many wan and faded cheeks
Have kindled into health?

The old, by thee revived, have said
"Another year is ours!"
And way-worn wanderers, poorly fed,
Have smiled upon thy flowers.

—*Wordsworth.*

ORIGIN AND GROWTH OF SOUTHPORT.

SOUTHPORT is situated on that part of the coast of Lancashire which lies between the estuary of the Mersey and the mouth of the Ribble, at a distance of about eighteen miles from the entrance to the former. Throughout almost its whole extent, the sea-border here presents a continuous range of sandhills, upon the outer or western side of which there is a broad belt of level sand covered with water at high tide, but left bare during a considerable portion of every twenty-four hours. Inland from the sandhills the country is flat for a distance of several miles, but then rises with very agreeable undulation, the highest points commanding fine views, and memorable as having been in by-gone days the locality of beacons.

The exact geographical position of Southport is 53° 38′ 40″ north latitude, and 2° 59′ 45″ west longitude.

The railway distances are:—from Liverpool, 18 miles; from Wigan, 17 miles; from Bolton, 27 miles; and from Manchester, 37 miles. Preston, 18 miles distant, gives ready access to Southport from Yorkshire and the North, and no places upon the north-western coast of England are more readily reached from the midland counties and the metropolis, either by Manchester, or by Crewe and Wigan. The nearest market town is Ormskirk, nine miles to the east. The nearest patrician mansions and residences, are Lathom House, the seat of the Earl of Lathom; Rufford Hall; Scarisbrick Hall; and Blythe Hall.

Relatively to the village of Churchtown, two miles distant to the north, Southport is of almost recent origin. At the beginning of the present century, it had scarcely come into existence. To imagine what was then the complexion of the ground now occupied by Lord-street and the Promenade, we must imagine the Birkdale sandhills continued uninterruptedly to a point beyond the Hesketh Park, with pools of water, and forests of sharp and rushy grass; narrow pathways leading crossways, at long intervals, to the wastes beyond. The cottages of a few fishermen stood about half-a-mile inland from high-water mark. Of other houses there were none. Churchtown, in a word, was the seat of local residence. Churchtown was the little mother village of modern, handsome, and wealthy Southport, and whoever would trace the uprise of the latter, must commence his survey with the immediate neighbourhood of the Churchtown Botanic Gardens.

The parish has for its name North Meols. What may be the precise etymological signification of the latter word is not clear. Baines, in the History of Lancashire, says it is a Saxon term, signifying sandhills.

Quite a hundred years ago, probably for an undeterminable earlier time, the sands of North Meols were noted for the facilities they offered to invalids who looked to pure air and sea-bathing as the chief means towards restoration to health. The physicians of Manchester and other manufacturing towns in South Lancashire, were accustomed to send their patients hither. Churchtown was the primary destination, and thence any who desired to dip in the sea were conveyed, when the tide served, to the suitable localities then called the Hawes.

The reputation of the neighbourhood grew fast. The wealth and the population of the cotton districts were constantly augmenting. The visitors to Churchtown became more numerous every season. Travelling two or three miles from their lodgings before the water could be reached, and this over rough and unformed sandy roads, was found to be a serious inconvenience, and in these circumstances Southport may legitimately be said to have had its origin. Some far-seeing and enterprising man is generally at hand to play the part of Columbus, though obscurely, when courageous adventure offers promise of reward. About the year 1792, a pioneer arose in the person of a Mr. William Sutton, landlord of one of the two inns then existing at Churchtown. To the amazement, and it would appear the amusement of his neighbours, Mr. Sutton erected, on the spot at the Birkdale extremity of Lord-street where the ornamental lamp-post now stands, a hostelry and little lodging-house, which he called the "King's

Arms." The original building was constructed entirely of wood. Before long, it was enlarged with more substantial materials, and in time became the "Royal Hotel" of the period. The little cabin has long since disappeared. The fact remains, nevertheless, that in this simple manner, Southport had its birth. The enterprise was considered so visionary that the "King's Arms" received, like many other buildings of the kind, the facetious appellation of "Duke's Folly," the "Duke" being the jocular title by which Mr. Sutton was distinguished. Poor man, he was doomed to experience the usual fate of those who live and work in advance of their times. His subsequent history was infelicitous, if not unfortunate. Although living to see Southport a thriving village, he died, without partaking of the prosperity around him, in 1841. The man who bestowed upon Southport its name, given, perhaps, because of its geographical position with regard to Churchtown, was a Mr. Barton, a retired Ormskirk surgeon, one of the most devoted of the early believers in the salubrity of the place, and who chose, as the fitting time of the bestowal, an entertainment given by Mr. Sutton, when his inn was first opened to the public. Notwithstanding the doubts that were entertained as to the prudence, not to say the sanity, of Mr. Sutton, it soon became obvious that he had taken the first step in providing for a genuine want, and that he had met it himself to only a limited degree. Cottages soon began to multiply in the neighbourhood of the "King's Arms"; and although these were intended at first for the accommodation purely of visitors, they soon became tempting residences with people who, coming as invalids, found Southport their best permanent abode. New and independent residences of superior quality soon followed, and, as in most other places of similar history, the growth of Southport has been steadily progressive. How wonderful the expansion in not exceeding eighty years, may be judged from the fact, that the last census returns (1881), give for Southport and Birkdale in the aggregate, a population of 42,454.

The second step as regarded the providing hotel accommodation was made about the year 1807, when the "Union" was erected. In this year was also built the first actual row of houses. In 1818 the Wellington Buildings made their appearance, and two years afterwards Southport enriched itself with its first ecclesiastical edifice, namely, Christ Church, Lord-street, now a beautiful structure with tower, lofty spire, and a peal of eight bells, but then of humbler appearance. Up to that time episcopal service was celebrated only at Churchtown, to which place even Birkdale still belongs, ecclesiastically.

Birkdale is the southern of the two townships into which the parish of North Meols is divided. There is nothing to indicate where one township ends and the other begins. The boundaries are not marked by a stream or other natural feature. The path from one extreme of the parish to the other is virtually continuous.

An important move towards the formation of the future town was made in 1825, when an Act of Parliament was obtained by the Lords of the Manor, authorising, among other things, the laying out of the present Lord-street. An avenue more noble is not often seen except in a metropolis. It runs almost exactly parallel with the high-water mark of the sea, about 400 yards to the east, and has now grown to be 1440 yards in length. Being of excellent width, the vista is either way exceedingly fine, and every year sees some addition to the very handsome buildings and frontages which confer its architectural character. In some parts this noble street is lined with villa residences, having little gardens in front, with abundance of trees. The greater portion, however, is devoted to commercial purposes, and towards the centre are found most of the municipal and other public buildings. A mass of stonework more imposing than that which comprises the Town-hall, the Cambridge Hall, the Art Gallery, the Post-office, and accessory edifices, it would be difficult to find in any city in the country. Of late years, great pains have been taken to convert large portions of Lord-street into a boulevard. Vigorous young trees have been planted, and in another quarter of a century, when these trees shall have attained fair dimensions, Lord-street, Southport, will certainly claim to be an object deserving of national admiration.

Parallel with Lord-street is the Marine Promenade, which, including the recent extensions north and south, now measures no less than 2600 yards. It was commenced in 1834, with a length of about 400 yards, reaching from the end of Nevill-street to Coronation-walk. In less than twenty-five years the length was considerably more than doubled, and great as it has now become, the probabilities are that it will be continued still further towards the north. Raised well above the level of the sands; admirably defended from the assaults of the tide, when unusually high and vehement; of capital width; excellently paved and asphalted; and bordered along the inward side by handsome terraces, with hotels and other large buildings, few sea-side roadways of the kind supply a more inviting place for exercise. It has the immense advantage also, of being open to the full influence of the afternoon and evening sunshine, the latter very generally implying sunsets of singular brilliancy. The coast of Lancashire has always been noted for the beauty of its sunset views, and these are certainly nowhere obtained to greater advantage than at Southport. Across the water, looking westwards, the eye catches the bold yet softly beautiful outlines of the mountains of the nearer portions of North Wales. At the other extreme of the sea-view, the picturesque sandhill range which marks the neighbourhood of Lytham, is plainly seen; and beyond this loom the grand heights of Cumberland and Westmoreland, including the celebrated eminence called Black Combe, the summit of which is 1919 feet above the sea level. In particular states of the atmosphere, the peaks of the Isle of Man are said to be distinguishable. These delightful views and prospects are obtained to even greater satisfaction from the Pier—one of the most remarkable in England. The great distance to which the water retires when "out," and which it was desired to neutralise, so as to allow of the approach of boats and steamers at any time of day or night, demanded a length of not less than 1465 yards. In addition to this, the extremity is expanded into a platform of 180 feet in length, and proportionately wide, so as to permit of the gathering together of many people, and give room for all. In order to facilitate approach to the extremity, whoever does not care to walk the whole distance has the option of a little tramway worked by a stationary engine. The original work was commenced in 1859. The entire structure was widened and lengthened in 1864.

When the water is out, the broad firm sands, which in parts stretch like a Sahara, are incomparable for horse exercise. Their substantial character, and the certainty of doing no mischief, recommended them for the early experiments made with the Whitworth long-range artillery. At the present day they are resorted to, periodically, for drill and other military exercises.

Returning to Lord-street, it is impossible not to be arrested by the pile of noble public buildings above referred to, and which may now be spoken of in detail.

The original edifice was the old Town-hall, erected in 1853, and fairly classical in its columns and pediment. Adjoining it is the still more handsome Cambridge Hall, the foundation stone of which was laid in 1872 by the Duchess of Teck. It was opened in 1874 by Sir R. Cross. The total cost was £30,000. The very elegant tower at the south front corner, with illuminated clock and a peal of bells, at once attracts the eye of every visitor to the town. The interior contains many spacious and well-proportioned apartments, and is partly occupied by the Post-office. The great hall is capable of holding 2,000 persons. Contiguous, in turn, to the Cambridge Hall are the Atkinson Art Gallery and the Atkinson Free Library, erected in 1877, at a cost of £8,000, which large sum was wholly provided by the munificent gentleman whose name is attached to these most valuable elements of the wealth of Southport—the late Mr. William Atkinson, originally a Manchester merchant. To this gentleman the town is indebted also for the clock-tower of the Cambridge Hall and the stone front and spire of Christ Church. The architecture of the exterior of the Free Library and Art Gallery, designed by the Messrs. Waddington, of Burnley, is composite Italian, a style which allows of the introduction of abundance of graceful ornament, without temptation to unwise luxury, and certainly a better example of chaste adhesion to the laws of purity in architecture would be hard to find. The main gallery of the portion devoted to the Fine Arts is 67 feet in length; two other picture galleries are each about 45 feet in length; and in addition to these there are smaller ones for sculpture and miscellaneous articles. The wall space available for pictures exceeds 6,000 feet. The Free Library contains, in the reference department, about 1,127 volumes, and in the lending department about 10,931 volumes. Other apartments are available for meetings, such as those of the Southport Literary and Philosophical Society, which usually assembles here.

Over and above its indebtedness to Mr. Atkinson, Southport has good reason to be grateful to another gentleman for some buildings, and for several scientific gifts of real beauty and usefulness — namely, the late Mr. John Fernley. Trinity Wesleyan Chapel; the School for the Daughters of Wesleyan Ministers, opposite to the Chapel—both exceedingly handsome structures; the Drinking Fountain, and Barometer upon the Promenade; and the Observatory in Hesketh Park were all provided by the munificence of Mr. Fernley.

In Lord-street are also found the Winter Gardens, an establishment so complex and complete in the variety of the enjoyment it provides—alike for the permanent inhabitants of

Southport and for visitors—that it may unhesitatingly be pronounced unique. The Brighton and Westminster Aquariums, which come nearest, have no actual gardens outside; here, on the other hand, the external portion is not inferior to any part of what is covered in. The area of this fine property is about nine acres. Possession of the ground was obtained about 1872 by a Company whose first expenditure approached £100,000, and who certainly selected a site which it would be impossible to consider other than the very best for such a purpose, having its front entrance in the principal thoroughfare of the town, tramway cars from north and south passing the gates every few minutes. Looking out in the rear upon the sea, the outside flower-gardens protected from high winds by a peculiar local depression of the ground they occupy, what locality more eligible could be found? Crossing the bit of garden space just within the gates, and entering by the principal doorway, the steps upon the left hand lead into the Aquarium, which is one of the completest in the kingdom, containing some thirty or forty tanks, with pools also for sea-lions and other large aquatic creatures. Upward steps from the same point lead into a splendid Pavilion, 170 feet in length and 44 in breadth, and which gives access at one extreme to the great Concert Hall, fitted to hold an audience of 2,000, and at the other to the Conservatory. When in 1805, Sir James Edward Smith wrote his notice of the evening-primrose for the original *English Botany*, he said his specimens came from some dreary sands a few miles north of Liverpool; they were gathered at a place without a name. Now, upon the very spot he speaks of, or close by, there is the noblest and loftiest botanical palace—after Kew and Chatsworth—existing anywhere in England. It is filled, moreover, with valuable and curious plants, including many of the most admired varieties of exotic evergreens, such as tree-ferns; and, in regard to its flowers, presents throughout the year an unrelaxing current of cheerful beauty. The length is 180 feet, the width 80 feet; and along the centre—which is an exact miniature of the transept of the Crystal Palace at Sydenham—the height is the same, namely, 80 feet. There is nothing else of the kind in Lancashire of similar altitude, so that such plants as palms have at Southport the best chance the county affords of a prolonged and beautiful existence. The Concert Hall is devoted to the purpose for which it was designed, and to the regular exhibition of dramatic entertainments. There is also a large Skating Rink.

Whilst upon the subject of provision for high-class pastime and recreation, the time is opportune to speak of the Hesketh Public Park, reached by tram-car in a few minutes from the northern extremity of Lord-street. Up till 1866, the site was chiefly occupied by sandhills. The thirty acres which are now so thoroughly covered with trees, shrubs, grass, and flower-borders, were then given to the town, as a present in perpetuity, by one of the principal landowners, the late Rev. Charles Hesketh, rector of North Meols. In 1868, the Park, with its encircling carriage way, was thrown open, and from that time forwards, it has been regarded as one of the most charming possessions of all who care for a lovely and peaceful resort, reached at a minimum expenditure of time. For the same reason, the present becomes the suitable opportunity for speaking of the Churchtown Botanical Gardens. That they are situated rather more than two miles distance from the centre of the town is no objection to their being considered as part of Southport. Two lines of tramcars go to the gates every ten or fifteen minutes; they are readily reached also by the West Lancashire Railway. If the Lord-street Winter Gardens show what enterprise, guided by taste, and with no stint of money, can accomplish, how splendid again the proof afforded at Churchtown! Up to about 1873, the locality was meadow and swamp, threaded by a little water-course. Twenty acres of this were enclosed; the stream was made the parent of a beautiful lake; by the constant employment of the curve, in laying out the ground, the paths were made so to diverge and glide into one another, that the twenty acres seem fifty; every portion was plentifully and judiciously planted with shrubs and herbaceous perennials, and to crown the whole, a magnificent Conservatory was erected, which in the north of England has scarcely a rival. Taking the three establishments together—the Winter Gardens, the Hesketh Park, and the Churchtown Gardens—the provision in Southport for the enjoyment of all who take delight in flowers, trees, the song of wild birds, and sweet fresh air, where there can be no fear of molestation, is beyond all dispute, in this part of England, unapproached. It should be added that the Churchtown establishment includes also a very fair Museum, chiefly ornithological. Upon the lake, moreover, there are boats for those who like rowing.

The Southport Glaciarium, at the northern end of Lord-street, has the distinction of being the only place in Great Britain, if not in the world, where skating on real ice can be enjoyed all the year round. The promoters—Mr. Edward

Holden at the head—have shown extraordinary courage and perseverance in overcoming the great difficulties that had to be encountered before the place could be said to be in perfect working order. They have now succeeded not only in providing a really dry, firm, and beautiful sheet of ice, 54 yards long, and 18 yards wide, but also in rendering the air of the building dry and pure by a ventilating process recently patented. The Scotch national game of curling, a most invigorating and wholesome exercise, has received an impetus by the opportunities given at the Glaciarium, and twice a year, clubs come from all parts of the kingdom to compete for a challenge shield, value 50 guineas, given by Mr. Holden. Pure block ice is also made and supplied in any quantities.

Southport contains plenty of places of worship. Christ Church, the original, as already mentioned, is now only the eldest of nine belonging to the Establishment, viz.: Holy Trinity, in Manchester-road, erected in 1837; St. Paul's, St. Paul's-square; St. Paul's School-Church, Duke-street; St. Andrew's, Eastbank-street; St. Luke's, Hawkshead-street; All Saints', Queen's-road; All Saints' School-Church, Ormskirk-road; and St. Philip's, Scarisbrick New Road. At a little further distance from the middle of the town, but in the borough, there are St. Cuthbert's, Churchtown—the old parish church of North Meols; and St. John's, Crossens. The Birkdale churches are St. James, Lulworth-road; St. Peter, Liverpool-road; and the Church School-room, Birkdale Common.

St. Cuthbert's, Churchtown, above named, has the peculiar interest attaching to it of being the only local memento of the past which can strictly be called archæological. Ecclesiastically, the place was in the first instance under the priory of Penwortham, near Preston. The church appears to have been built at three different periods, the earliest assigned date being 1571. It stands upon the site of a still older church supposed to date from the twelfth century. The earliest existing records are a tombstone in the graveyard, dated 1577; and a register for 1594. Inside the building there are three marble monuments of interest: one, the work of Nollekens—greatly admired for its artistic qualities—to a member of the Hesketh family. The same family is commemorated in the second piece of sculpture; the third was erected to the memory of Thomas Fleetwood, of Bank, who died in 1717. The Fleetwoods (an old Staffordshire family) were the original patrons, Edward Fleetwood presenting in 1684. The connection of the Fleetwood and Hesketh families, as well known, is of long standing. The memory of the late rector of North Meols, the Rev. Charles Hesketh, above-mentioned, is widely cherished, alike on account of his Christian character, his benevolence to the poor, and the excellent use he made of his varied opportunities as the rector of the parish for 44 years, and as an extensive landowner in the district. The present rector is the Rev. C. Hesketh Knowlys.

The Wesleyans have excellent chapels in Mornington-road, Leyland-road, Duke-street, and Upper Aughton-road, Birkdale—the majority of them very handsome. The Congregationalists are quite as well provided for in Chapel-street, Portland-street, and Lord-street West. The Catholics have a pretty building, dedicated to St. Marie, in Seabank-road, erected in 1841 from designs by the elder Pugin, and enlarged in 1875; with a second in Albert-road, Birkdale, dedicated to St. Joseph. Other denominations are represented to the extent of at least seven or eight; the total number of places of worship counting up to fifty-five. A census taken on a Sunday in September, 1882, showed the presence of over 13,000 worshippers in the morning, and 12,300 in the evening, at the different places of worship in Southport and Birkdale. These numbers show that, in proportion to the population, there is probably no community in the country distinguished for a more becoming and respectful Sunday observance. At a census taken in 1851, Southport established a reputation in this respect which it has never lost, the percentage of those attending Divine service being found to be 87·8 of the population.

The Cemetery, opened in 1865, lies upon the south-west side of the town.

Southport is well off, likewise, in respect of its Market. Up till 1880 the dealers in "fish, flesh, and fowl," fruits and vegetables, had their stalls in a large building in Chapel-street. This being insufficient for the constantly increasing needs of the town, a new one, very airy and commodious, was erected close by, at a cost of £40,000, and formally opened by the Earl of Derby, September 7th, 1881. Upon this day also the New Promenade Extension was opened with much ceremony by the Earl of Lathom.

The Banking fraternity are well represented, having five branch establishments, viz., the Preston Bank; the Manchester and Salford Bank; the Manchester and Liverpool District Bank; Parr's Bank; and the Southport and West Lancashire Bank. Some of the buildings occupied are very ornate, especially those of the last-named Company, which stand at the corner of the group of municipal buildings mentioned above, and form a strikingly handsome addition to them.

One of the most important and interesting of the Southport institutions has still to be mentioned—the Convalescent Hospital and Sea-bathing Infirmary. The origin of this very useful and well-conducted establishment is found far back in the local history. In the year 1806 a charitable fund was organised by some benevolent persons, with a view to enabling the poor of the large manufacturing towns in Lancashire and Yorkshire to receive, when invalided, the benefits of sea-air and sea-bathing. This fund, small no doubt at first, was steadily augmented from without as the excellence of the design became apparent. The appellation given to the institution—probably the pioneer of its kind—was the "Strangers' Charity." In 1853 the present building was erected. In 1862 it was considerably enlarged, and the name was at the same time altered to that one now in use. New buildings, of very considerable beauty and extent, are now in progress of erection, these additions being rendered practicable through the liberality of the committee of the "Surplus Cotton Famine Fund," which has contributed for the purpose a sum of no less than £40,000. The number of beds available for patients is at present 220, and when the extensions shall be completed there will be accommodation for 400. The institution is now the second in the kingdom in point of extent, being surpassed only by the Margate Royal Sea-bathing Infirmary. To all appearance it will before long stand in advance of all, a blessing to the operative classes—especially of the manufacturing districts—which it is impossible to over estimate. Numbered among its active supporters have been many of the chief philanthropists of the large towns of Lancashire. It is aided also by the great landowners and other principal capitalists of the county, the clergy and ministers of all denominations also doing good service. During the year 1860 the number of patients received was 1454; last year (1882) the number was 2628. The Convalescent Hospital has a good supply of salt-water for the indoor use of its inmates, being connected, by a line of pipes, with the Victoria Baths, which draw it continually from the sea by means of powerful engines.

The Victoria Baths constitute one of the most valuable possessions of Southport. They were erected in 1871 at a cost of £40,000. Externally, the building is allowed to be very handsome, the style of architecture being Italian classical, and the frontage extending to a length of nearly 200 feet. Facing the sea, and being central, the edifice is a great ornament to the Promenade. The interior arrangements are excellent. It may be allowed to claim for them that they are the best in the kingdom. Every description of bath is provided—salt water and fresh water, hot air and vapour baths, are alike at the command of the public. Visitors to Southport and invalids have thus a permanent choice in regard to place for sea-water bathing. Any difficulties or disadvantages that may at times arise in regard to bathing upon the open sands are perfectly met and compensated within the walls of this very admirable establishment.

Southport, like other towns, has its Infirmary and local charities. There are also some excellent institutions, not so general in character, which deserve special mention, viz., the Convalescent Home, in Lord-street West, in connection with the Manchester and Salford Provident Society; the North of

England Convalescent Home for Children, in Hawkshead-street; the Governesses' Home, in Alexandra-road; and the Home for Gentlewomen, in Park-road.

Though not among the very first of the Lancashire towns to secure to itself the advantages of the railway system, Southport was not long behind. The line from Liverpool to Preston, viâ Ormskirk and Burscough Junction, opened April 2nd, 1849, gave easy access to the north, and a choice of ways to Yorkshire. The line to Liverpool, viâ Crosby and Waterloo, was opened October 1st, 1850. Through communication with Manchester was established, viâ Wigan, April 7th, 1855. A direct line to Preston, called the "West Lancashire," was opened September 4th, 1882; this line also provides another route to Blackburn. A new line, called the Southport and Cheshire Lines Extension Railway, connecting Southport with the Midland system, viâ Aintree, is in course of rapid construction, and will be opened in part before the close of the present year (1883).

The government of Southport is vested in a mayor, ten aldermen, and thirty councillors. The Charter of Incorporation was issued in 1867. The first election took place June 1st, of that year. The burgess roll contains 4,891 names.

The public Press has been well represented for a long course of years, though with varying fortunes to the different papers. The oldest local newspaper—still pursuing "the even tenor of its way"—is the *Southport Visiter*, the first number of which was issued in 1844. The other papers now in course of issue are the *Southport News* and the *Southport Guardian*.

Southport has long been renowned for the number and

excellence of its Schools. It has been computed that the number of pupils in the schools, sent hither from other places, is at present (1883) about two thousand. In 1869, Southport was made a centre for the Cambridge University Local Examinations, and has distinguished itself as a centre by the marked success of the boys and girls examined here every December. It has been far above the average success of other centres. This is an excellent testimony to the thorough education given in the schools of Southport. The total number of youths and girls examined since 1869 has been 1,359, of whom 945, or upwards of 69 per cent., have successfully passed the Examinations. Upwards of 27 per cent. of these Examinees have passed, moreover, in honours. At the Examination in December, 1882, the average of the Southport candidates who came up to the Examiners' requirements was nearly 13 per cent. higher than the average of all England; and, in comparison with the results attained at some of the great city centres, Southport appears in a still more favourable light.

The political franchise is held by 2,361 persons—1670 in the borough of Southport, and 691 in Birkdale. The constituency forms a portion of the south-western division of Lancashire.

The land upon which the town is built is owned by representatives of the Hesketh and Scarisbrick families. A large portion of Birkdale is the property of Mr. T. Weld-Blundell.

The rise in the value of public and private property has been commensurate with the increase of the population, the grand total being now about £12,000,000. The average annual rent of the houses is £40.

The Southport Waterworks Company obtained their first

Act in 1854. They provide for the town and district a good supply of pure water, drawn from the New Red Sandstone at two pumping stations, one at Aughton, the other at Town Green.

In 1878 the town was greatly benefited by the introduction of a new system of drainage, which, though hardly yet complete, has proved, as the vital statistics show, a great success. The Corporation, recognising the importance of sanitary improvements, have spent upon these drainage works £150,000. The outlet is into the sea at Crossens, four miles to the north of Southport.

It needs only to add that the principal hotels in the town—many of them very large and handsome—are the Victoria, the Royal, and the Queen's, on the Promenade; the Prince of Wales, the Bold Arms, and the Scarisbrick, in Lord-street.

In Birkdale there are two large Hydropathic Establishments, viz., "Smedley's," Trafalgar-road; and the Palace Hotel and Spa. The Company owning the last-named has recently erected large and handsome baths of various kinds, supplied with salt water direct from the sea.

CHAPTER II.

When the extent of benefit which may be derived from occasional change of air, both to the physical and moral constitution, is duly estimated, no person whose circumstances permit will neglect to avail himself of it.—SIR JAMES CLARK.

SOUTHPORT AS A RESORT FOR INVALIDS.

GEOLOGY OF THE DISTRICT—GENERAL REMARKS ON CLIMATE—
LOCAL CLIMATE OF SOUTHPORT.

GEOLOGICALLY considered, Southport is situated upon the edge of a series of recent deposits overlying the margin of the Trias or New Red Sandstone of the south of Lancashire. The rocks of the latter form the high ground towards Ormskirk and Liverpool. Nearly the whole of the space to which the name of Southport, with its immediate adjuncts, may be applied, was, as regards the surface, at no very remote period, a tract simply of blown sand; whatever soil may exist upon any part of the surface within these bounds has been created by the industry of man, and but for cultivation the greater part would have remained sand to the present day, containing only a trace of organic matter. The space in question begins at Waterloo, and extends almost unbrokenly, though with varying width, to Crossens, Southport thus occupying what is nearly the north-western extremity. The greatest width is at Formby, where the tract of blown sand is nearly three miles across. The narrowest part is at a point which would be represented by a line drawn from the mouth of the river Alt to Orrel-hill Wood. The extent of the whole is from 14,000 to 15,000 acres. The bed of sand varies considerably in depth, thinning out inland, and undergoing every possible superficial change, some of the higher dunes rising to as great an elevation as seventy-five feet above the mean level of the sea. The average thickness of the entire deposit has been estimated at about twelve feet. Underneath the blown sand there are variously arranged beds of peat and silt, red loam, boulder clay, and laminated blue clay, or siliceous silt; while here and there occur intercalated beds of soil which have been at some period under cultivation but have again been covered up. Sections in the courses of streams, and in artificial openings, often indicate as much as four or five feet of peat. The surface of these peat-beds appears to dip towards the sea, and near the coast the covering of sand is usually three or four feet in depth. At Formby the peat crops out upon the shore, and here, also, the presence of numerous stumps of trees certify the ancient presence of a forest. The very low level of the peat near the sea declares pretty certainly that in past ages there has been a considerable amount of local subsidence. The result of this would be that the drainage would be obstructed, and thus, that the ancient forest would gradually disappear. There seems to be no evidence of the subsidence extending beyond the boundary of the alluvial deposits. The peat-beds, as already said, dip towards the sea, and in some places are actually below the line of the rise of the spring tides, so that a sinking or contraction of the underlying sands seems to be the only cause to which the phenomenon can be attributed. At a little distance inland, the peat-beds are exposed upon the surface of the ground, and here the peat is collected and dried for use as cottage fuel.

Some light was thrown upon the character of the deeper strata by an unsuccessful attempt made a few years ago to obtain water in Birkdale by boring. At sixty-five yards depth, beds of red and light coloured marls, with crystals of sulphate of lime and white granular gypsum, were met with; at sixty-seven yards, similar beds, with greater quantities of sulphate of lime and gypsum; at sixty-nine yards, beds, again similar, but without sulphate of lime and gypsum. All the deposits contained common salt, which was very evident on the application of the tongue. Similar red and variegated marls, containing gypsum, have been met with in the upper Permian beds of Lancashire, and in the keuper marls of the trias in Cheshire, but in the latter alone has salt yet been found. It is probable, therefore, that the Birkdale deposits are triassic, and that the strata is not very likely to yield an abundant supply of fresh water.

There can be no doubt that the British Islands were subject, in the primeval times, to the rigours of a Polar climate. Great Britain was once only a scattered archipelago of murky, misty islets, their chief phenomena the mighty, crushing glacier, and the electrical flashings of Boreal lights. By what means, and through what processes, they were brought to their present state is in some measure open to conjecture, the most probable being changes in ancient sea levels, and the establishment of strongly defined ocean currents from the mass of heated water around the equator.

The position of the British Islands on the map, and the unusually mild temperature they enjoy, are so inconsistent that it might puzzle the theorist, as well as the practical man, if he attempted to explain the fact without taking into calculation the above all-important cosmical influences. Within a few degrees of the region of perpetual snow, Britain has an atmosphere equal to that of any part of the temperate zone. To tell the nervous, the consumptive, or the hypochondriac, that they are living in a country about the same distance from the Arctic Circle as are the inhabitants of Labrador and Kamtschatka would, at the least, add an extra chill; to tell the delicate invalid, hastening to the sheltered coast of Devonshire, that he is fixing his winter dwelling to the northward of the latitude of the Banks of Newfoundland, would certainly impart an unwelcome shock to his sensibilities. But isothermal lines are not coincident with parallels of latitude, and the modifying circumstances of climate do more than correct the evils of position—they very often introduce a bland and salubrious element in situations of the most unpromising description.

The insular position of these islands, after all, would be of little avail had they not some more vital bond of union with more genial climes. That magic "circle of marriage with all nations" would form but an icy bond were there not some currents bringing life and warmth to our coasts. The extent of influence possessed by the great Gulf Stream in these climatic modifications is not yet fully understood; but there can be no doubt that it has had a large share in the changes to which reference has been made.

A nation and its destiny may be linked by very slender threads. Should any deep, mysterious, but all-potent cause, ever throw those mighty activities into new and unaccustomed channels, thereby producing startling changes of local climate, the historian's fancy sketch of the meditative New Zealander may be realised by means of agencies of which he never dreamed.

The effect of proximity to the sea in softening and warming a climate has long been recognised. Owing to the penetrability of water by radiant heat, and the perpetual agitation and intermixture of its superficial strata, its changes of temperature are neither so extensive nor so sudden as those of the land. An island is always found to possess a milder air than land in the same parallel of latitude forming part of a continent. On this subject, Sir Charles Lyell well observes:— "The ocean has a tendency to preserve everywhere a mean temperature, which it communicates to the contiguous land, so that it tempers the climate, moderating alike an excess of heat and cold."

In addition to this general result of proximity to great masses of the ocean, some localities derive additional benefit from certain great marine currents which transport their waters from about the equator.

It is generally believed that the water encircling our shores during the winter months is some degrees warmer than our atmosphere. It is also considered that the west coast of our island is milder than the east coast. Why this should be so is not easily explained without bringing into the question the qualifying influences supposed to be derived from the currents of heated water setting from the equator. Neither to the winds that blow, nor to the sun that shines, can these differences be wholly referred. It is not a theory, but a positive fact, that a portion of the Gulf Stream impinges on the west coast of Ireland, bearing abundant traces on its bosom, in the shape of fragments of tropical vegetation, of the hot latitude from which it has travelled. It is, indeed, possible that these shores would enjoy a milder climate than they do at present, did they not contribute to the sea so many large rivers fresh from the hills, serving to put a barrier of cold water round our shores, and absorbing the surplus heat from the warm currents. Many significant data might be procured if proper observations were made at such parts of our coast as are favourable to the required experiments. Many of our fashionable summer resorts on the coast might be found to owe the peculiar mildness of their climate to the proximity of these currents, and to the absence of any great outlet of fresh water into the sea.

Strong sea currents, setting over the south coast of Ireland, would be deflected towards the estuary of the Mersey, and as the currents of the Dee and Mersey prevent them ascending those channels, these waters would be pressed towards the north, and may tend to raise the temperature on the Southport coast.

The fact has been proved, that, while the deep sea water

in the channel remains of an average temperature, that of the flood tidal water, coming over the banks, is higher than either the sea or the air. It gives out its surplus temperature, and is probably one cause of the mildness and salubrity of Southport in the winter.

At the end of this book will be found a series of meteorological tables, compiled by Mr. Joseph Baxendell, from observations made at the Southport Meteorological Observatory, Hesketh-park, during the eleven years 1872-1882. They form a complete and valuable record of the rainfall, temperature, and humidity of the district.

Careful examination of these tables will show that Southport enjoys a remarkably equable climate, and that the climatic advantages formerly claimed inferentially, are actually found to exist, when subjected to the test of rigorous daily observation, made with duly verified instruments.

The sandy nature of the soil enables the moderate rainfall to be easily disposed of, and, as there is reason to believe that the evaporation in this district is much in excess of the rainfall, it is apparent how dry the average surface must be, and how much the mean humidity of the atmosphere is modified.

The nature of the soil tends to equalize the temperature as well as to elevate it; the first of these effects being of chief importance, as it is not always a *high* temperature which is most desirable in the climate of a sanatorium, but an *even* temperature, which is neither too hot in summer nor too severe in winter. This is precisely the character of the mean temperature as observed here, and gives Southport an immense advantage over places much further south, which many persons would think are, therefore, warmer. Comparisons have shown that the mean temperature of Southport is less variable than that of some well-known health resorts on the south coast, taking each quarter of the year as the basis of comparison; and that, although the position—200 miles more to the south—raises the mean temperature for the year one degree, the temperature fluctuates more than that of this locality. Certain portions of the Welsh coast enjoy the influence of the Gulf Stream more than Southport, and, hence, surpass even the south coast of England in mean annual temperature; but this advantage disappears when we go more into detail, and compare the mean daily ranges of temperature of these places. This is the true criterion of climate, so far as it depends upon temperature only, and, tried by this test, the climate of Southport is much more genial than that of many other places in lower latitudes. Climate, however, is not a mere question of temperature. Many other elements enter into it, and, notably, humidity. Here, also, the condition to be desired is that of regularity; neither excessively dry air nor excessively moist air. Comparisons of mean humidity show in a striking manner the advantageous character of the climate of Southport, as regards the more equable condition of this important meteorological element. The prevailing winds, as shown by Mr. Baxendell's observations, are from the south and the west. A seaside place has, of course, a higher rate of wind movement than obtains inland, but the other meteorological conditions being singularly equable, and the lighter winds prevailing, it is, perhaps, an unmixed advantage for the purification of the atmosphere that the total movement of the wind should be as high as it is.

The above general remarks upon the circumstances which modify all climates, and upon the local peculiarities of Southport, must be borne in mind in proceeding to consider the claims which this place possesses as a resort for invalids. Sharing, as it does, with the most favoured health resorts, the advantages derived from immediate proximity to the sea, it has certain well-marked and more immediately local advantages, which few other places possess in an equal degree.

Foremost amongst these must be placed its open sea aspect, yet well-sheltered position on the coast. It has all the advantages enjoyed by other neighbouring watering-places, without the exposure to bleak and piercing winds incident to some towns on the north-west coast. East and north-east winds are usually limited to the months of April and May. The tide receding a considerable distance leaves a large expanse of sand to be heated by the sun, which has the effect of warming the sea-breeze passing over it, giving to Southport, that of which few, if any, other watering-places can boast, viz., a bracing sea atmosphere, and yet, one *thoroughly dry*. Whether the air immediately in contact with the sand, while parting with its moisture, does not not take up some of the peculiar constituents only found in sea water, or whether a stratum of dry air passing over an extended sandy surface, at a high velocity, has not its force of electrical tension highly increased and condensed, are questions worthy of consideration. Both have been adduced as explanatory of the peculiar sanitary effects of Southport in some diseases.

In addition to the abundance of ordinary oxygen in the atmosphere, there is in the Southport air another form of the same element, to which has been given the name of ozone. This substance, as the etymology of the word suggests, is known by its peculiar smell, which somewhat resembles that of burning sulphur. The odour may be perceived in a room in which electrical or galvanic machines have been in action, and often in the open air after a thunderstorm. Experimentally, ozone may be produced by passing electric sparks through confined portions of air. Its presence, in association with lightning, is thus easily explained. Ozone (discovered by Dr. Schönbein) is oxygen in an allotropic condition, three volumes of oxygen forming two of ozone. Hence it derives the peculiar combining power which ordinary oxygen does not possess. Whether the presence of ozone be due to the electrical state of the atmosphere, to partial decomposition of water, or to the presence of peroxide of hydrogen, it is usually most plentiful upon the surface of the sea, and to it the invigorating power of the sea air is in a great measure due. Indirectly, ozone is beneficial in destroying noxious effluvia and miasms. Indeed, it may be regarded as nature's disinfectant, accomplishing or accelerating the oxidation of all decomposing animal and vegetable matter, and being a potent agent for the destruction of the germs of zymotic disease. Owing to its energy as an oxidizing agent it is often difficult to detect in the air of large towns, and wherever there is much decomposing animal matter. As ozone is found most plentifully above the surface of or near the sea, and where vegetation is scanty, and is associated, moreover, with the prevalence of south-westerly winds, it is not surprising that it should constitute a distinguishing feature of the air of Southport.

The usual mode of gauging the amount of ozone present in the air is to expose test slips made of blotting paper soaked in a solution of iodide of potassium and starch. The slips are suspended in a little cage, roofed in so as to shelter them from the rain and the direct rays of the sun, and the degree of discoloration that takes place in a definite time is then carefully noted and compared with a fixed scale. The results of such tests are, however, not trustworthy, as no account is taken of the varying wind force, a most important factor in making a correct estimate. Besides, nitric acid, peroxide of hydrogen, and other agents occasionally present in the atmosphere, may exert a similar action upon the iodide of potassium, and be thus partly or wholly responsible for the production of the iodide of starch colour. Mr. Baxendell gave some interesting lectures in 1881 upon the results of his tests for ozone, extending over nine years. They were made at the Observatory in Hesketh Park, where he had the advantage of comparing the wind force and direction at the same time. His tables show that the amount of ozone in the Southport air is large, and also that it has increased since the sanitary condition of the town has been improved.

Southport having two water lines, at a great distance apart, the climate of each has its own distinctive quality; that at high-water mark having all the characters of the stronger and more stimulating one at low-water, but in a more modified and milder form. Beyond this inner line, and more in the line of the streets of the town, the sea breeze is found still more softened, the atmosphere is buoyant and remarkably free from impurity and humidity. Invalids being able to avail themselves of the varied qualities of these distinct climates according to their changing condition, is of great practical importance in the treatment of disease. Nor should the patient himself neglect to study and observe these differences, which, though apparently trifling, are capable of helping or retarding the progress of his case. An injudicious walk on the Promenade in cold weather has often undone the work of weeks; whilst on the other hand, from the want of suitable guidance, the period of convalescence has been needlessly prolonged from an undue fear of exposure to a bracing atmosphere.

The character of the soil and of the surrounding country adds greatly to the sanitary value of Southport. The soil, consisting chiefly of sand, retains no moisture or rain upon its surface, a heavy fall of rain leaving no trace after a very short time. The fall, indeed, is slight in comparison with that of the adjacent country, which, being more hilly, attracts the rain clouds more readily. The absence, in its immediate vicinity, of any considerable body of fresh water, is another climatic advantage, very few places having the same extent of country free from running or stagnant water. The facility thus afforded for taking exercise is of the utmost importance to invalids. In localities situated upon the clay, a heavy shower, for even a short time, involves the suspension of outdoor exercise for a day or two, in consequence of wet roads and atmospheric evaporation. Under such circumstances, which of course are of frequent occurrence, delicate people are either compelled to encounter the risks attendant upon wet feet and breathing a damp air, or are precluded from the muscular exercise upon which depends the healthy condition of all the animal functions.

The atmosphere of Southport is remarkably free from malarious influences; epidemics rarely occur, and when they do they are seldom malignant. It exerts upon visitors a sedative and composing influence.

It is proper to remark here that notwithstanding the truth of the above statement as to the dryness of the air, it is not so excessive as to be irritating to the skin or mucous surfaces. Such a condition would be nearly as injurious as the opposite extreme of excessive humidity. Even during the prevalence of the east and north-east winds—those most unpopular of the subjects of Boreas—the irritative effect is not greater than is found in other localities; much less indeed than upon the east coast of England, whilst the prevailing westerly winds come softened by the vast expanse of the ocean.

Besides the consideration of meteorological data and vital statistics, there is a mode of determining the curative influence of climate of not less importance, and which has been too much overlooked; that is the effect of any given climate upon the native population. By observing the peculiar nature of the climate, and its influence upon the stationary inhabitants, assistance is given to discriminate also in the choice of cases of disease likely to be benefited by being sent to such a locality. If in any climate, it is found that the agency is decidedly of a relaxing kind, and that it proximately acts by modifying the tone of organs, *à priori*, it would be inferred that such a climate is unsuitable to that kind of diseased action depending upon general want of tone, and a low state of functional energy. But, again, if in any climate acute inflammatory affections—for instance, of the mucous membranes of the air passages—are a common disease with the natives, it would not seem to be a wise or logical proceeding on the part of a physician to send to such a climate a person who was likely to be affected by these very maladies. Now, if these principles are applied to the case of Southport, everything advanced in favour of its climate will meet with the fullest confirmation. That the climate is at once bracing and sedative, may be gathered from the physical and moral history of the population. If we take as a type of these, the fishermen, we find them broad and fleshy in their frames, phlegmatic in temperament, slow in their movements, and (though this must be attributed to something better even than a good climate) remarkably decorous and staid in their conduct. Amongst the natives, we also find many cases of extreme longevity.

Here may be quoted the picture of an imaginary climate for the consumptive, as drawn by the eminent physician, Dr. W. B. Richardson, leaving such readers as are acquainted with Southport to judge how far it meets the case.

"I shall recommend no particular place as a resort for consumptives, for I wish not to enter into disputation on this point. But there is a formula for an hypothetical consumptive Atlantis. It should be near the sea-coast, and sheltered from northerly winds; the soil should be dry; the drinking water pure; the mean temperature about 60°, with a range of not more than ten or fifteen degrees on either side. It is not easy to fix any degree of humidity; but extremes of dryness or of moisture are alike injurious. A town where the residences are isolated and scattered about, and where drainage and cleanliness are attended to, is much preferable to one where the houses are closely packed, however small its population may be."

CHAPTER III.

See the wretch, that long has toss'd
   On the thorny bed of pain,
At length repair his vigour lost,
   And breathe and walk again.
The meanest floweret of the vale,
The simplest note that swells the gale,
The common sun, the air, the skies,
To him are opening paradise.

GRAY.

EFFECTS OF THE CLIMATE UPON DISEASE.

GENERAL CLAIMS OF SOUTHPORT AS A SANATORIUM.

SUGGESTIONS FOR INVALIDS.

IN offering some advice to those who have left home in pursuit of health, it is necessary to dwell somewhat at large upon the importance of maintaining a hopeful state of mind. Though it is said,

    "Hope springs eternal in the human breast,"

it is not easy to cherish and retain that feeling under circumstances of declining strength, of long continued or oft returning pain, and isolation from all the habits and excitements of accustomed duties. The nervous depression which chronic illness naturally induces, often leads an invalid to take a more gloomy view of his condition than the facts will justify. Of course there are cases where a reasonable hope of recovery can no longer be entertained; and in all cases of protracted illness it is the duty of a Christian to prepare for the most solemn issue, that it may be also the most welcome and most blessed. But there are special reasons, derived from the inherent powers of the system, and amply confirmed by experience, which afford sufficient ground for a chastened hope, even in circumstances of undoubted gravity. The chief illustrations of this are found in connection with one of the most formidable complaints which afflict humanity—Consumption. Pathological facts show that recovery from consumption may take place, and there is conclusive evidence that tubercle does occasionally become absorbed.

These evidences of the fact of recovery in consumption, are found in cases where death has occurred from other diseases; but we have equally valid testimony during the life of some who have been its subjects. There are many who have presented all the rational signs or symptoms of consumptive disease, and every year adds to the number. Many have recovered from the first stage, and, doubtless, more such cases would be recorded if the nature of the complaint were better appreciated by the public, and earlier attention paid to declining health, previous to the appearance of special chest symptoms. A large number of instances of the arrest of this disease have been made known.

It must not be supposed that successful attempts of nature to check the progress of this formidable complaint are of rare occurrence. As an instance of the life-protracting influence of modern therapeutic agents, it may be mentioned that Dr. J. B. Williams—than whom no man is better qualified to speak on the point—asserts that the average duration of consumption, formerly estimated at two years, may, under improved treatment, be fixed at four years. If these things be so, and we are entitled to entertain a reasonable share of hope, even in the case of so formidable a disease as consumption, with how much greater propriety may this be done in most other complaints? Advanced life, in connection with disease, affords less ground for hope; but in early and middle life, we do well to have faith in the reparative powers of Nature, assisted by the resources of art, especially when the system has not been undermined by a previous career of debilitating excesses.

A change of residence from a humid atmosphere to a mild dry one promotes the equable distribution of the circulating fluids over the whole system, increases the activity of the capillaries of the surface, and in the same proportion diminishes the congestion of internal organs. The continued action of a bland atmosphere upon the delicate surfaces of the respiratory tubes, lessens irritation and assists in the more efficient production of those changes of the blood so essential to health. These are sufficient reasons to account for the importance of change as a means of recovery in various forms of illness. The hope engendered by a new movement taken towards recovery; the cessation of business cares and anxieties, novel scenery, new associations, and the other incidents attendant upon a change of residence,—all have a powerful effect upon the weakened frame. And when the locality chosen is appropriate to the particular ailment under which the patient actually labours, or with which he is threatened, and especially when the measure has been taken in an early stage of the complaint, the result is often of the most valuable kind, and justifies all that has been said by those who place change of air among the foremost of our remedial agents.

The maladies which change of climate is most likely to help to alleviate are chronic bronchitis, asthma, emphysema, strumous diseases, consumption, chronic rheumatism, chronic dyspepsia, ulceration of the fauces, clergyman's sore throat, some forms of paralysis, and nervous depression after long illness; and to these complaints the climate of Southport is especially adapted. The advantages of a prolonged residence on this part of the coast, in connection with the more immediate treatment of symptoms, are such as arise from its marine position, and from the constant operation of its peculiar local climate.

The most direct and certain remedy for many chronic sufferers, is the habitual breathing of an air containing a maximum amount of oxygen. The proportion of the constituents of atmospheric air remain nearly the same on the highest mountain as in the deepest vale, the principal difference being the amount of carbonic acid mixed with it in different localities.

Owing to the pressure of the superincumbent atmosphere, air increases in density the nearer we approach the level of the sea, and it is evident that we inhale at every breath a greater amount of air, and, consequently, a greater amount of oxygen, than at a few hundred feet higher. One great secret, therefore, of the cure of chronic cases at the sea-side, is the being able, without extra exertion or effort, to receive into the lungs an additional amount of oxygen. The effect of this is to rouse and sustain the nervous system, and to expedite and perfect the aëration of the blood in the lungs, by means of the more rapid combustion of carbon, thus creating a greater demand for nourishment, as shown by the vigorous appetite which so generally follows a removal to the sea-side.

As might be expected from what has been already stated, the climate of Southport is peculiarly adapted to the prevention or relief of consumption. In the earlier stages, particularly, before tubercles have actually formed, the effects are often most surprising. The prolonged residence here of young persons threatened with this fearful malady, has in numerous instances perfectly re-established their health; or in the case of those who possess an hereditary tendency to the disease, has postponed the accession of fatal illness. When the lung has been more or less affected by tubercular deposit, the favourable conditions found in this climate have often, with very little medical interference, arrested the progress of the mischief; and, by giving every advantage to the great restorer, Nature, have induced a marked diminution of cough and expectoration, the gaining of flesh, and the return of bodily and mental strength.

The same results follow in many cases of chronic bronchitis, attended by excessive secretion and exalted sensibility of the pulmonary mucous membrane. The relief in these cases, brought about by a change from a cold and moist to a mild and dry climate, especially when aided by a judicious use of some of the preparations of iron, is, perhaps, more marked, because often more rapid than in any other morbid condition. It may be stated, in general terms, that the same external circumstances that prove advantageous in consumption, are of equal value in this complaint.

Decidedly beneficial results are witnessed also, in emphysema of the lungs; the tonic and sedative effects of the atmosphere exert a favourable influence upon the air passages, reducing the secretion, improving the breathing, restoring sleep, and, these ends attained, the general health gradually and surely improves.

In internal congestions, pulmonary in particular, in heart diseases, asthma, and indeed whenever there is imperfect circulation of the blood, or difficulty of breathing, the extreme *purity* of the Southport air is found to add greatly to the comfort of the invalid, and where the disease is of short standing, and circumstances are favourable, residence in this locality is highly conducive to a cure. In the aged, in whom there is reason to believe that structural change has already taken place, disease has apparently stood still for years, and a degree of comfort has been experienced to which the patient has long been a stranger while living on a clay soil, or in the neighbourhood of copious vegetation. Elderly people suffering from asthma generally find the air of Southport suitable for them, and many have made the place their permanent residence for this reason; it is not uncommon to hear them say that they cannot breathe so well anywhere else.

In chronic rheumatism, and general, or partial paralysis, the recovery is frequently very remarkable. That it should be so in the former case will be understood, when we remember the dryness of the atmosphere, and the injurious effects of damp upon sufferers from rheumatism. The relief of paralysis is probably due, not only to the improvement of the general health, but to the reduced pressure upon the nervous centres, arising from a light and pure atmosphere.

The importance of so pure an atmosphere, possessing such physical peculiarities, in diseases of a more general nature, is sufficiently obvious. The unwholesome conditions to which the dwellers in pent-up cities, and unhealthy districts, are habitually exposed, lead to the production of a low tone of the general health, and proclivity to disease, rendering them very susceptible to prevailing epidemic influence. The comparative freedom from epidemics hitherto enjoyed by the inhabitants of Southport, affords the best illustration of the converse of this truth.

The climate of this place, in conjunction with sea-water bathing, has a peculiarly beneficial effect in certain forms of cutaneous affections, which are extremely distressing to the patient, and are often among the least satisfactory cases with which the physician has to deal. Among these may be mentioned acne, psoriasis, lepra, and troublesome chronic eczema. The capillary vessels partake of the improved tone communicated to the system at large, while the sub-acute inflammation of the skin is at once soothed and subdued by the application of the sea-water.

In the large variety of diseases comprised under the general term of scrofula, a lengthened residence by the seaside is acknowledged to be by far the most important means of cure. The number of young children with feeble, ricketty frames, ulcerating glandular enlargements, and pallid countenances, is lamentably large. The local complaints under which they suffer, are only the symptoms of constitutional degeneration, which requires the long-continued employment of constitutional measures for removal or improvement. A marine atmosphere, sea-bathing, warm clothing, nourishing diet, and other hygienic measures are the essential remedies. The special advantages which Southport offers in these cases, over other sea-side resorts, are the dryness of its atmosphere, and its walks, the safety of its sea-bathing, and the unfailing occupation which children find in digging in the sand.

It might be supposed that the advantages of a sea-side residence, as well as the other special local advantages offered by Southport, would be of little importance in affections of the digestive organs. Such a supposition, however, would be erroneous. Many forms of dyspepsia are greatly relieved by change from a raw cold climate to a warmer locality, in conjunction with the utmost attention to diet, and regular exercise, either on horseback or on foot. In those cases of dyspepsia particularly, where the mucous membrane of the stomach is irritable, the improvement is very marked. The same may be said of similar states of the intestinal membrane, in chronic diarrhœa. It would be impossible to particularise the affections of the liver and other organs which have been benefited by this climate, or which, at all events, have seemed to owe their cure to a long continuance of its influence.

The forms of dyspepsia which seem to derive most benefit from the climate of Southport, are those which present—in addition to the usual local symptoms—an enfeebled and languid condition of all the functions, a pale countenance, the body emaciated, the extremities cold, the skin harsh and dry, the intellectual faculties impaired, and the muscular force diminished, so that mental and bodily exertion are equally difficult. These symptoms, which are continually presenting themselves, seldom fail of relief, if the sufferer will pay a moderate attention to diet, exercise, clothing, and to those general sanitary rules which have been a thousand times repeated, and need no further reiteration.

A few words of caution are necessary, with particular reference to those who suffer from affections of the throat and chest. Although, as already shown, there are few days in which an invalid cannot contrive to get walking exercise at Southport, it must be mentioned that the changes of temperature during the same day are frequently considerable. It is needful, therefore, carefully to avoid going out either too early or too late in the day. During certain portions of the winter, not more than two or three hours intervene between the chills of morning and of evening, and this interval should be chosen for taking out-door exercise. It is also desirable, indeed absolutely necessary, in more serious cases, that the patient should keep his rooms at an equable temperature, say of about 60°, and this should be done both by day and by night. The great and sudden change from a warm sitting-room to a cold bed-room, is continually frustrating the best contrived attempts to bring about a cure.

Few things are of more importance in the management of chronic disease than that a rational and well-considered plan of treatment should be pursued with perseverance, and for a sufficiently lengthened period. And yet the anxieties of the invalid frequently lead him to err on this point. Not reflecting that his present condition has been the result of a long continued divergence from the standard of health, in some one or more of the functions or organs, before there resulted what forms his actual disease; he forgets, or does not understand, that the healing powers of Nature, however encouraged and aided by Art, when they have really begun to remedy the evil, can only return to the healthy condition at a similar pace. He lays himself open, consequently, to every promising offer of a royal road to recovery. Systems surround him on every side promising the speedy fulfilment of his most ardent wishes, their claims endorsed by this and that enthusiastic friend. Comparisons are made between his case and others, based upon the slightest resemblances, and without even an attempt to ascertain how far those resemblances are real or only apparent.

The mingling of truth with falsehood which exists in medical heresies, is the real source of their success. A system of pure error could not exist for a day. But when a portion of truth is recognised in an otherwise false system, it conceals its real nature as a whole, owing to the difficulty of discriminating in matters so alien to an invalid's ordinary pursuits. But it must be admitted that the present state of medical heresies is to some extent a legacy from the former system of medical practice, and which, possibly, has still its adherents. It is only fair to say that the present state of things cannot be altogether accounted for by the weakness and credulity of the public; something must be put down to the mystery and excessive medication of former times. The public were greatly to blame for the mystery, since they persisted in attributing a power to the medical man beyond all reason;

they were to blame also for an undue use of medicine, since they supposed that in medicine alone consisted his power to do them good; and if one practitioner declined to prescribe, they went to another. But still the profession were consenting parties. There was a want of confidence in the force of truth, when urged with simple earnestness. Had the profession been sufficiently alive to the danger of reaction in the public mind; had they calculated upon the growing intelligence of society; had they sacrificed their immediate interests to the permanent welfare of the profession, they would have prevented the present discreditable state of things. We are not now speaking of vulgar quackery: that must always exist while the masses are ignorant and unreflecting, and thus in danger of becoming the prey of designing men. We allude to those fashionable systems which are followed by so many otherwise thoughtful and intelligent men and women, who are not to be led astray by mere credulity, but require some one guiding principle, of which they must be persuaded. This has been with many the conviction that the former practice of over-drugging with medicine was wrong. Satisfied of this fact, they have dwelt upon the discovered truth so long as to have little thought to expend upon the foundations of the system they have adopted. They know themselves to be right on one point of the enquiry, and they too lightly assume the correctness of the rest. Tired of so much physic, they fix upon water, a remedial agent of good repute, and erect a temple of health in which she is the exclusive goddess. As hydropaths, they can, at least theoretically, get rid of the drugs they so much detest. Or, if unprepared absolutely and ostensibly, to "throw physic to the dogs," they tamper with their reason so far as to substitute a semblance for a reality, and, having minutely subdivided the "dummy," swallow it with the greatest possible gravity. Prove to them, if they will listen,—which they will seldom consent to do,—that their fundamental principle is a falsehood; remind them that for the production of every positive effect there is required an exactly adequate cause; show them that their great conclusive arguments, their reputed cures, are but prime examples of the logic of *post hoc, ergo propter hoc*, and that the same syllogism would equally establish all the competing systems of quackery that now exist, or have ever existed; do all this, and more, yet they fall back upon their first strong persuasion, and behind that entrenchment stand till events prove to them the fallacy into which a partial truth has led them.

There is one point, not bearing exclusively upon the condition of the actual invalid, but of more general interest, to which allusion may be made, that is the subject of prophylactic medicine, or the department which has reference to the prevention of disease. That this department should have received so little attention is indeed surprising. It is a popular saying that "prevention is better than cure," but patients and physicians have alike been content to leave the matter in its proverbial form, so far as any systematic carrying out of the principle is concerned. Very scanty notices of this subject are to be found, and those are widely dispersed, in medical writings. It is so much the custom virtually to limit the duty of the physician to the cure of disease that this noble sphere for the exercise of his skill and ingenuity is practically ignored. And yet it is probable that, in a large proportion of those who die of chronic disease, the seeds of such disease have been implanted by the time they have attained their fortieth year. Would it not be wiser to make the first rudimentary appearance of anything in the shape of local or general derangement into a *casus belli*, the ground of a regular attack, rather than to wait till offensive hostilities appear in the form of painful symptoms? An unwonted sensation, or a marked change of function, amounting in neither case to positive inconvenience or distress, may, nevertheless, be significant of approaching ill, since we know that here also, "coming events cast their shadows before." It is reasonable to suppose that suitable antidotal means might often be devised, based upon the physiological changes going on, to prevent those structural alterations which are sure to follow abnormal action longcontinued. This, however, can only be called prophylactic in an accommodated sense; but we would go further, and urge the necessity of a true prophylaxis. The transmission of hereditary tendencies to disease is of constant occurrence; individual peculiarities are often attended by a proclivity towards certain forms of physical derangement; a misguided early training may have warped the frame in an evil direction; certain employments or modes of life lead, without fail, to injurious and well-known results. All these, and many others that might be mentioned, are instances in which a careful system of preventive measures, not taken up and applied intermittingly, but dovetailed, so to speak, into the economy of life, would seem to be the dictate of true wisdom. People are so much in the habit of thinking that men must die of disease, that a healthful old age is looked upon as remarkable, something for the attainment of which no special effort can be made. No legitimate object of human desire can fail of at least partial accomplishment where proper means are properly brought to bear upon it; and yet few would be found to contend either that a healthy longevity is not such a legitimate object, or that it is not generally left to the merest hap-hazard.

No better instance can be given of what a due attention to prophylactic means can accomplish than the case so well described by Dr. Watson, in his admirable Lectures on the Principles and Practice of Physic: "The late Dr. Gregory, of Edinburgh, used always to mention in his lectures the case of Dr. Adam Ferguson, the celebrated historian, as affording one of the strongest illustrations he ever met with of the benefit that may be derived from timely attention to the avoidance of those circumstances which tend to produce plethora and apoplexy. It is, perhaps, the most striking case of the kind on record. Dr. Ferguson experienced several attacks of temporary blindness some time before he had a stroke of palsy, and he did not take these hints so readily as he should have done. He observed that, while he was delivering a lecture to his class, the papers before him would disappear—vanish from his sight, and appear again in a few seconds. He was a man of full habit, at one time corpulent and very ruddy; and though by no means intemperate, he lived fully. I say he did not attend to these admonitions, and at length, in the sixtieth year of his age, he suffered a decided shock of paralysis. He recovered, however, and from that period, under the advice of his friend, Dr. Black, became a strict Pythagorean in his diet, eating nothing but vegetables, and drinking only water or milk. He got rid of every paralytic symptom, became even robust and muscular for a man of his time of life, and died in full possession of his mental faculties at the advanced age of ninety-three, upwards of thirty years after his first attack." Sir Walter Scott describes him as having been, "long after his eightieth year, one of the most striking old men it was possible to look at. His firm step and ruddy cheek contrasted agreeably and unexpectedly with his silver locks; and the dress he wore, much resembling that of the Flemish peasant, gave an air of peculiarity to his whole figure. In his conversation, the mixture of original thinking with high moral feeling and extensive learning, his love of country, contempt of luxury, and especially the strong subjection of his passions and feelings to the dominion of his reason, made him, perhaps, the most striking example of the Stoic philosopher which could be seen in modern days."

But immoral indulgence of the passions and appetites, and the more obvious infractions of the physical laws, with the neglect of wise precautionary measures, are not the only points upon which it is needful to take warning. The intellectual and emotional nature of man is subject to laws quite as stringent as those which regulate his bodily functions. The injurious influence of mental excess is not less positive than that of physical, though not so obvious. It may be difficult to persuade the busy man on 'Change that the growing dyspeptic symptoms which trouble him are the direct result of the state of turmoil to which his brain has been exposed for months and years together; and yet the fact is certain. The student of law or divinity who strains his faculties to the utmost, without allowing them the repose necessary to recruit them, is not only sinning against his own body, but is adopting the best plan to thwart his own cherished objects. The popular minister, whose whole soul is in his work, and who is compelled to keep his intellectual powers on full stretch to meet the requirements of his position, while his life is passed in a succession of nervous excitements, exposed to alternations of heated rooms and cold night air, is undoubtedly doing a great work, but he does it at a great cost. He will hardly live to build up the Church by his matured wisdom, or exhibit the passive virtues of the aged Christian. The list of highly gifted ministers of various Churches who have been lost to mankind when in the full vigour of their intellectual and moral strength, by a systematic neglect of the most ordinary sanitary rules, is sad to contemplate. The subject is one of great delicacy, and it is only necessary to suggest that the moral government of God being perfectly harmonious in all its parts, the fulfilment of a duty in one direction never necessitates opposition to the Divine intention in another.

Intellectual labour, pursued in the quiet of the study, if too long continued, and not sufficiently alternated with out-door exercise, is fertile of ill effects. The maladies thus induced are extremely varied, and not seldom are attributed to any cause but the right one. They may take the form of a direct injury to the over-worked organ, the brain, and may proceed onward along the parallel lines which lead respectively to insanity or paralysis. But more generally they will assume one of the protean forms of dyspepsia, and lead to impaired nutrition or structural change. Sydenham considered that one of the most severe fits of gout he ever experienced, arose from great mental labour in composing his treatise on that disease; and the student of literary history will call to mind many instances where the completion of some intellectual masterpiece has been speedily followed by the death of the master. The late gifted Hugh Miller is one of the many examples of this fact. It is to be lamented, that those who "intermeddle with all knowledge," and who are the appointed instructors of mankind, should so often neglect that knowledge with which their own mental and physical comfort is closely connected, and the acquisition of which would multiply their capabilities of usefulness to the race.

If prophylactic measures have an important bearing upon the subject of the prolongation of life, not less important is the proper treatment of advancing age. Although an individual may escape destruction from causes that are accidental and extraneous, he nevertheless bears about him natural and internal causes of decay, inevitable in their progress, and leading to one certain result. With the germs of life are intermixed the seeds of death; and, however vigorous the growth of the bodily frame, however energetic the endowments of its maturity, we know that its days are numbered. To mark the gradual succession of the phenomena which attend these changes is deeply interesting. In youth, all the powers of the system are in excess of its demands, and the body increases in bulk. In course of time, the processes of reparation and decay approach nearer to an equality, and at length are exactly balanced. By a wonderful system of adjustments the balance is kept perfect, often for many years, until, at last, old age steals on by slow and imperceptible degrees. The relative proportions of the fluids and solids are altered, the solid tissues become condensed, muscular substance appears almost changed into tendon, fibrous structures either lose their flexibility and become too rigid for use, or are changed into bone.

The smaller arteries are obliterated, and the heart undergoes structural change; functions are feebly performed, the chemical condition of both solids and fluids becomes altered, the skin grows dark and corrugated; and, as the various signs of decay increase,—the tottering step, the bent form, and the palsied movement,—we perceive that the individual has entered upon that period, when, in the sublime language of Scripture, "The keepers of the house shall tremble, and the strong men shall bow themselves, and the grinders cease because they are few, and those that look out of the windows be darkened, and the doors shall be shut in the streets, when the sound of the grinding is low, and he shall rise up at the voice of the bird, and all the daughters of music shall be brought low; also when they shall be afraid of that which is high, and fears shall be in the way, and the almond tree shall flourish, and the grasshopper shall be a burden, and desire shall fail; because man goeth to his long home, and the mourners go about the streets: or ever the silver cord be loosed, or the golden bowl be broken, or the pitcher be broken at the fountain, or the wheel broken at the cistern. Then shall the dust return to the earth as it was; and the spirit shall return unto God who gave it."

When and how this descent towards the tomb shall take place, is in the hands of Him who measures out our days, and appoints our outgoings and incomings. Human science is impotent in presence of the general evidences of decay. But where the stress of disease is so localised as to threaten destruction before these marks of decay have become general, she can sometimes relieve that stress; she can suggest the compensations required by altered circumstances; she can endeavour to remove the obstinacy which persists in retaining habits no longer applicable or safe; she can erect barriers against anticipated evils; she can soothe the irritability of weakness, and assuage the violence of pain. At all events, her ministers can never be more legitimately employed than in the struggle to prolong human life; and their efforts will be more or less effective, in proportion to the attention they may give, not only to actual disease, as it affects the different periods of life, but also to its first incipient manifestations. And it is to this dawning stage of illness, before the evil has attained any considerable power, that the attention of those whom it concerns should be invited. Practical effect should be given to the maxim—" Prevention is better than cure."

CHAPTER IV.

> This is the purest exercise of health,
> The kind refresher of the summer heats;
> * * * *
> Even from the body's purity, the mind
> Receives a secret sympathetic aid.
>
> THOMSON.

ON SEA-BATHING.

THE importance of bathing as a hygienic and therapeutic agent has been recognised by all nations, at all periods of history; its practice existed as well amongst nations basking under the heat of a tropical sun, as amongst the hardy inhabitants of the unthawed regions of the north. By the former it was employed as a religious observance or mode of luxury, by the latter with a view to health, or to counteract the effects of intense cold.

The histories of Greece and Rome furnish abundant evidence of the extent to which bathing was practised by these nations. So fascinating to them was the luxury of the bath that it was customary to employ it at their festive entertainments, and it was considered essential to the *eclât* of public rejoicings. Establishments for this purpose were constructed, vieing with each other in magnitude and splendour, as may be seen from the ruins which still excite the wonder and admiration of the traveller.

The importance of bathing cannot be overrated if we consider that the skin upon which it operates performs the several functions of absorption, secretion, and excretion, and that upon its surface the bloodvessels and nerves terminate. It has also a wide range of sympathies, in which are included the alimentary canal and air passages, and it co-operates also with those great emunctories of the circulating system, the lungs, the liver, and kidneys, aiding them in the elimination of noxious matters. Hence, the absolute necessity that there should be no impediment to the performance of its functions.

Sea-bathing has many advantages over ordinary bathing. The sea may be considered practically as a medicated bath, containing, besides well-known saline constituents, iodine and bromine in minute proportions, which latter exert a peculiar action upon the glandular and absorbent system. The sea is also the habitation of innumerable organic beings, who live and die there; it therefore becomes impregnated with subtle and volatile animal particles, which extraordinarily increase the stimulating powers of sea water. We conclude, therefore, that open sea-bathing, where it can be borne by the invalid, is preferable, as in home or in-door bathing, although all the elements of sea water may be present, there is still the absence of a saline atmosphere, of the shock of the waves, the agitation of the water, and the electric and magnetic currents which are evolved, and exert a stimulating effect upon the system. It will be well to enlarge a little on these topics.

Sea-bathing on the British coasts (for its action is very different in the tropical waters of a warm climate) owes its efficiency to the combined influences of *cold*, of the *saline particles*, which enter into the composition of sea-water, and of the *shock* produced by the impulsion of the waves. In order to understand its effects we must endeavour to form a just estimate of the power of each one of these agents separately. The first impression produced by the cool temperature of the sea, which even in summer rarely exceeds 67°, is powerfully to stimulate the numerous sensitive nerves of the skin. As all our organs are under the influence and direction of the nerves, every part of the body must therefore be excited and stimulated by the sea-bath; as when a bell is struck, the vibration extends over every part of the metal. Sea-bathing goes far beyond the mere local action on the skin, its immediate effect being a general stimulation of the whole nervous system. The sudden application of cold to the surface is followed by a shrinking of the skin and contraction of the tissues. As the result of this, the capacity of the bloodvessels is diminished, and a portion of their contents suddenly thrown upon the internal organs. Hence follows the participation by the nervous system in this sudden congestion, causing a more energetic action of the heart, and consequent rush back to the surface. This is the state termed *reaction*—the first and final purpose of every form of cold bathing. Reaction is known by the redness of surface, the glow and thrill of comfort and warmth, which follow the bath.

By it the internal organs are relieved, respiration is lightened, the heart is made to beat calmly and freely, the mind feels clear, the tone of the muscular system is increased, the appetite is sharpened, and the whole organism feels invigorated.

The stimulating effects of the *saline constituents* in sea-water form the second agent acting remedially. These, which constitute about one fifty-fifth part of its weight, produce a powerful stimulant effect upon the skin, and determine a more copious flow of blood to that organ, assisting the primary reaction, and shortening and diminishing its depressing effect. Owing to these qualities of sea-water, one may bathe in the sea at a lower temperature than in fresh water. Reaction, even in robust constitutions, is much longer in making its appearance after bathing in rivers; but in the sea, even on a calm day, and to a weakened constitution, it is almost instantaneous, and much more powerful. It has been supposed by some that the absorption by the skin of a portion of the saline ingredients may tend to increase these effects.

To illustrate the influence of the third element in a sea bath, viz., the shock produced by the *impulsion of the waves*, we need only refer to the effects of a douche bath to form an adequate idea of the difference experienced between bathing in a calm and in an agitated sea. The shock of the waves in a rough sea is, in fact, an extensive douche bath, which, by striking a great part of the body at once, makes all the more powerful impression upon the economy.

The general result of sea-bathing, both on the healthy and invalid subject, is to stimulate nutrition and improve the functions of every organ, increasing the vitality of the blood and improving the various secretions of the body. The action of the skin is augmented, the liver pours out a greater quantity of bile, and a more active respiration consumes a greater quantity of carbon. In consequence of this increased activity, the system gradually purifies itself of a mass of worn-out particles, which were tolerated so long as the body was in a languid state, but which, under the stimulus of increased energy, it casts off as an oppressive load. Thus we see the strengthening process giving rise to an alterative action in the diseased frame; swollen and indurated glands, scrofulous tumours, cutaneous eruptions, and other morbid deposits, are re-absorbed, and thrown out by the system.

There are certain conditions which require to be attended to, with regard to the differences of strength, constitution, and temperament, in individual cases. The first caution required is not to continue the immersion too long. Even in vigorous subjects, prolonged immersion is very apt to be followed by injurious effects, the danger being greater in proportion to the coldness of the bath. After the first shock on entering the water, a feeling of warmth and a genial glow is perceived; if the bather quits the water before this stage passes away, the whole surface of the body will partake of the sensation; if immersion be prolonged farther than this, the blood is driven to the internal organs, the nervous energy is depressed, and reaction being prevented, injurious consequences are liable to ensue.

One of the first of these is weakness of nervous energy, with irregularity of muscular contraction. No doubt most of the accidents that occur in bathing, and are generally referred to a supposed seizure of *cramp*, arise from this cause, viz, the enfeebling effect of undue cold upon vital action. This is perceived in the difficulty of fastening the dress when the hands are chilled. Hence persons of a spare and slender habit of body, even though they be good swimmers, should be cautious of venturing into deep water, especially at an early period of the season, when the water at the surface is no true indication of its temperature beneath. Even when the results of too long an immersion are not so directly injurious, the system suffers from other evidences of defective reaction, such as a sense of chilliness, which continues throughout the day. Though cold never injures the body when acting as a stimulant, yet, in delicate and convalescent persons, the sensations of the bather must be specially regarded in relation to its mode, duration, and degree. The time occupied in bathing in cold water by invalids, though varying according to individual cases, should not, as a general rule, exceed a few minutes, say from two to ten. Before entering the water, a smart walk should be taken along the shore, so as to produce a comfortable glow, and assist the reaction. Persons in moderate health may remain in the water a longer time, in this respect being governed by their own experience; but they must not omit the use of active exercise, both during and after the bath.

When the bather is suffering from nervous exhaustion from bodily fatigue, when the skin is cold and covered with moisture, or where there has been violent perspiration from the effects of medicine or exercise, the effect is sometimes to overpower the system rather than to rouse it to reaction. Care must also be taken not to allow too long a time to elapse in the preparation for the bath, and particularly not to hesitate too long before plunging into the water. It is in this cold stage that there may be danger, for the excitement has already passed away, and the system cannot resist the depressing influence of the cold. If the surface of the skin be dry, and the heat somewhat above the natural standard, little is to be feared from immersion into a lower temperature.

The next important question is the proper time for bathing. In delicate subjects, injury is frequently caused by cold bathing at a time when the vital powers are too languid to admit of the necessary reaction,—before a meal, or after any great fatigue, for example. The rule for the invalid should be, not to bathe either just before or just after taking food, nor after too long a walk. A bath early in the morning, before breakfast, exerts a more powerful effect than one taken at a later hour of the day, and requires proportionate energy and strength in the bather. As a general rule, both bathing and exercise, on an empty stomach, will be found unsuited to the invalid, and the best time will be the period between breakfast and dinner, taking care to avoid the other evil of bathing on a full stomach, which is dangerous to persons of full habits, or advanced in years, exposing them to the risk of congestion of the brain or even apoplexy. Two hours after breakfast and three hours after dinner should elapse before bathing is ventured upon.

Too frequent bathing is to be avoided. Bathing, like all other stimulants, depends principally upon its occasional use for its legitimate effects. The evils resulting from too frequent bathing are nearly equal to those resulting from too long immersion. The practice of bathing every day is not to be recommended. For persons of a delicate constitution and reduced habits of body, a bath every third or fourth day is sufficient; after a short period it may be tried every other day.

If the system be very weak and reduced, it is advisable to take a few preparatory warm sea-water baths, having the temperature daily reduced, so as to pave the way for bathing in the open sea; or a system of preliminary partial sponging with cold sea-water may be adopted, increasing the surface wetted daily, and commencing with the chest and back. As the good results of sea-bathing depend very materially upon securing the proper amount of reaction, where this is not attainable in the ordinary way, means should be used to bring it about; for this purpose the flesh-brush, or horse-hair gloves, may be used, both before and after the bath,—applying friction more particularly over the stomach, chest, and back. No doubt the best mode of using the bath is that of quick immersion. As cold bathing has a constant tendency to propel the blood towards the head, it ought to be a rule to wet that part as soon as possible; by due attention to this circumstance there is reason to believe that violent headaches might often be prevented.

There are many reasons why aged people should bathe with great caution. The tendency to disease of the brain increases as age advances, and it is very important that sudden and violent excitement be avoided; the strictest moderation should be maintained in every kind of mental and physical effort. In youth and manhood the waste resulting from the exercise of mind and body is soon repaired; but after the maturer years of life are passed, a point is reached when what is lost is lost for ever—any attempts to force either mind or body only leads to exhaustion. The warm bath is much more likely to be productive of good results in persons so situated, especially when the system is reduced from disease or over-exertion.

As a general rule, it may be said that wherever organic disease or change of structure exists, sea-bathing is injurious; debility, either nervous or muscular, being the type of those diseases in which it proves beneficial. As a practice, the most delicate as well as the most robust may be so trained as to enjoy and receive benefit from it; but there are some constitutions, more than others, which are liable to feel its ill effects. Such are those who are plethoric and of a bilious temperament, whose natural habit of body is to make blood rapidly. Where the venous and arterial systems are in a constant state of tension, sea-bathing would be found too stimulating a remedy, tending to produce a momentary congestion of blood in some parts of the body, thus producing unequal distribution, and a strain or pressure on certain organs. Of course the above remark applies more particularly to constitutions weakened by disease. Sea-bathing is no doubt enjoyed as much by persons of a full habit and bilious temperament as by others, and as safely, when properly trained to it.

Although the sea-bath is allowed to be useful in local congestion arising from debility and loss of vitality in an organ, yet even in these cases care and attention are required to prevent the weakened organs from becoming permanently injured by the quickened but unequal distribution of blood to the part. Individuals with a feeble action of the heart, or subject to spitting of blood, or in whom a state of active inflammation is present, should be particularly careful to use the bath with moderation, and to take advice before venturing on it.

As preparatory to, or instead of, bathing in the open sea, the warm sea-water bath is universally applicable. By its means invalids may gradually prepare themselves for the more stimulating and invigorating influences of the cold bath, who might not otherwise have been able to withstand the shock. Thus employed, it is better to diminish the temperature of the bath five or six degrees each time, trying the effect of applying cold to the back while immersed in the bath. Persons whose nerves are very irritable and cannot easily bear the shock of the first dip in cold water, and cannot bear the loss of animal heat, should not try the experiment, nor need they relinquish the good to be obtained by bathing. In the graduated scale of the temperate, tepid, and warm bath, a very little attention will enable them to hit the right medium, and they will thus possess an excellent substitute for the open sea.

Tepid and warm sea-water bathing has many uses. It acts as a sedative, promoting diaphoresis and determining from internal organs. It is advantageous also in nervous affections, rheumatism, gout, in certain cutaneous diseases, and in hepatic dyspepsia.

CHAPTER V.

How wond'rous is this scene ! where all is formed
With number, weight, and measure ! all designed
For some great end ! where not alone the plant
Of stately growth, the herb of glorious hue,
Or foodful substance ; not the labouring steed,
The herd and flocks that feed us, not the mine
That yields us stores for elegance and use ;
The sea that loads our tables, and conveys
The wanderer man from clime to clime ;
The rolling spheres that from on high shed down
Their kindly influence : not these alone
Which strike e'en eyes incurious ; but each moss,
Each shell, each crawling insect holds a rank
Important in the plan of Him who framed
This scale of beings ; holds a rank which lost
Would break the chain, and leave behind a gap
Which nature's self would rue.

STILLINGFLEET.

NATURAL HISTORY OF SOUTHPORT AND ITS ENVIRONS.

THE Natural History of Southport, surrounded as it is with sheer sand, extending inland for some miles, would appear to offer little variety in its objects, yet it possesses a *Fauna* by no means contemptible. Of Quadrupeds there are none but the ordinary little creatures familiar to everyone ; of Birds, an extensive variety ; of Reptiles, none of the family of snakes, but an abundance of other kinds ; of

Fishes, the variety is not great; of Insects, the number is considerable, including many that are esteemed rare. The list of Mollusks is a slender one, and the Shells found on these shores make no very great pretensions to diversity. Commercially regarded, the shore is in some respects immensely rich. So vast is the yield of cockles, that tons are frequently sent off at a time; and of shrimps the almost daily capture is enormous. Among the sandhills, at varying distances inland, there are shells of which we find no living representatives upon the shore, obviously deposited at some distant period, when the sea extended over a large tract of country now of considerable elevation.

In Botany, the plants common to uncultivated ground and marshy places near the sea-coast occur in profusion. Of Ferns there are but few; Mosses are numerous, and include several kinds which have hitherto been found only in the neighbourhood of Southport. Of other Cryptogamic plants there are plenty, so that at all seasons it is possible to procure botanical subjects of one kind or another.

THE SOUTHPORT FLORA.

Although destitute of the romantic scenery which usually implies corresponding variety of botanical habitat; although entirely wanting in deep and shady forest, limestone cliffs, waterfalls, and running streams; Southport is undeniably rich in curious and interesting wild-flowers. Of late years, it must be acknowledged, the great amount of draining carried on in the suburbs of the town with a view to adaptation of the land for building purposes, the construction of new roads, and the conversion of much of the originally wild surface into lawn, park, and garden, has tended to diminish the primitive abundance of the indigenous wildflowers; and in many localities, once noted for their plenty, the flora of half-a-century ago has entirely disappeared. It is difficult, however, for man to eradicate any plant that has once been well established in a given district. Not a single species can be supposed to be wholly lost from the Southport flora. Undisturbed localities still exist, a little further off it may be, but to be found by searching for; and very curious is it to observe how often, in the most highly cultivated spots, the aborigines re-appear, the seeds having been buried in the earth at depths too deep for them to vegetate, and biding their time until the spade of the gardener brings them near enough to the surface to be excited by the sunshine and the rain.

The peculiar physical geography of Southport gives the flora a distinctly two-fold character. First, there is the sand-hills section, including the maritime plants which grow within actual reach of the tide, and to which may be added the very interesting group of semi-palustral plants observable in the "slacks" among the sandhills—the depressed and often low-lying hollows which become perfectly dry only in the middle of the hottest summers. Secondly, there is the purely meadow, pasture, and inland wayside section, to which may be added the vegetation of the ponds and ditches, and that of the local piece of reclaimed moorland called "The Moss." The aggregate amounts, it would appear, to about 300 different species, or about a fifth of the entire number of flowering plants accounted indigenous to Great Britain.

Of native trees and shrubs the number is, of necessity, very small. Few, perhaps, can prefer the slightest claim to be considered truly wild, except the hawthorns and sallows in the inland hedgerows. Many, however, of the most admired and interesting ligneous plants of the country have been introduced, and these, in gardens, plantations, and other cultivated land, as a rule, thrive admirably, and give that very agreeable sense of leafiness which in sheltered situations is augmenting every day. Without reckoning the crowd of exotics now giving such excellent promise in the Hesketh Park and in the Churchtown Botanic Gardens, the variety and the capital complexion of the purely British kinds which have been brought together in those two admirable enclosures, show that the climate of Southport is congenial to an exceedingly high per centage. The only native tree which the Southport air and soil seem not to agree with is the yew.

The Southport sandhills form, without question, one of the most remarkable features of Lancashire. Continued southwards almost to Liverpool, the surface occupied by the whole has been calculated at not less than twenty-two square miles. The history of the original formation is a matter also into which the mind cannot but enquire with great interest. In Mr. Leo Grindon's "Lancashire" (1882), p. 54, it is stated that in the opinion of the distinguished geologist, Mr. T. Melland Reade, they have taken certainly not less than 2,500 years to acquire their present dimensions—probably a much longer time. "Some of the mounds, however, are palpably quite recent, interstratifications of cinders and matter thrown up from wrecks being found near the base. A strong westerly wind brings up the sand vehemently, and very curious then becomes the spectacle of its travel, which is like thin waves of transparent smoke over the flat, wet shore. The wind alternately heaps up the sand and disperses it, except where a firm hold has been obtained by the star-grass, which, running beneath the surface, binds and holds all together. A very beautiful decoration of the smooth surface of the declivities is constantly produced by the wind whirling the stalks half way round, and sometimes quite so when there is room for free play. Elegant circles and semi-circles are then grooved in the sloping sand, smaller ones often inside, as perfect as if drawn with compasses. Another curious result of the steady blowing of the sea-breeze is that on the level of the shore there are innumerable little cones of sand, originating in shells, or fragments of shells, which arrest the drifting particles, and are, in truth, rudiments of sandhills such as form the great rampart a little further in." The passage we quote illustrates exactly how in all probability they began. Some small object standing literally "in the teeth of the wind" would arrest the particles of drifting sand, and just as flowing water accumulates behind a barrier, in the course of centuries the tiny mound would swell into a hill, and valleys and ravines would follow as a matter of course.

The "star-grass" above-mentioned is one of the two specially characteristic plants of the sandhills, the other being the little salix which covers many portions with grey-leaved scrub. No sandy shore is devoid of it, but here it grows in patches so dense as often to resemble fields of corn, a likeness sustained by the flower-heads, which may be compared to ears of wheat. Many of the Southport people call it "maram," a word altered probably from the Danish "marhaulm," literally "sea-straw." That the Scandinavian voyagers and colonists of a thousand years ago left many traces of their visitings upon the coasts of Lancashire, has often been pointed out by antiquaries, and the presence of this old word would seem to supply another illustration of the ancient Danish presence. Botanically, this useful grass is the *Ammophila arundinacea*. The underground stems run to a length of many yards. Extending itself in this manner, advantage is taken of the centrifugal tendency of the plant to strengthen the sandhills artificially; portions of stem being pegged down wherever it is desired either to promote accumulation of the sandy particles, or to prevent wasting away.

The little grey salix, the foliage of which often shines with silvery lustre, is botanically, as the name imports, in essential characters, a willow. Very pretty, in early summer, are the innumerable catkins; and a few weeks later, when the ripe white cottony seed is discharged, most curious is the spectacle, the quantity being so vast as often to be gathered up by the eddying wind in what, but for the season, might be taken for snow-drifts. Several different varieties have been distinguished, but there are no absolute and permanent differential marks. It is quite enough to speak of the plant shortly and simply as *Salix repens*, the sandhills willow.

Upon the roots, or half-buried stems and branches, there occurs here and there, beyond Birkdale, and thenceforwards to Ainsdale, that extremely curious parasitic plant, the "yellow Birds-nest"—*Monotropa Hypopitys*. Like most other parasitic plants, it is entirely destitute of green leaves. Everything, from the ground upwards, is yellow. The stems, five or six inches high, are primrose-coloured; the nodding crest of flowers is primrose-coloured; and as if enamoured of consistency, the odour of these reminds one of cowslips.

Where the sand is not covered by maram or salix, and is fairly consolidated, we may often see the hound's-tongue, *Cynoglossum officinale.* The stems, which attain the height of about two feet, bear abundance of little purplish claret-coloured flowers. When gone, they are succeeded by great prickly seeds, growing in fours, and that catch hold of one's clothing below the knees, and like burs, refuse to let go.

In company with the hound's-tongue, there is plenty also of the Carline-thistle, *Carlina vulgaris,* prickly, like all the others of its race, but totally different in the colour of the flowers, which are yellowish and glossy, and remarkably sensitive to changes of the sky.

Not far from these will also be found two species of the very singular plants called "spurge,"—the *Euphorbia Paralias,* and the *Euphorbia Portlandica.* The first-named, the great sea-spurge, is two feet high, well-clothed with narrow leaves, the insignificant flowers greenish-yellow; the stem, very tough, filled with sticky milky juice, which owes its qualities to the presence of caoutchouc, or "India-rubber," too small in quantity, however, to be worth the trouble of extracting. The other species, the Portland spurge, is dwarf and bushy. The little roundish leaves assume in October most beautiful shades of amber and crimson. The juice, as in the Paralias, is milky and sticky.

Dotted about, still where the sandy slopes are quite open, there is a fair sprinkling also of Eryngo, *Eryngium maritimum,* the Touch-me-not of the sandhills. Every portion of the plant presents a *chevaux-de-frise* of strong and very pungent prickles;—unfortunate for those bent on gathering bouquets, since the flowers, produced in egg-shaped heads, are of the loveliest azure.

The Bugloss, *Lycopsis arvensis*, again attracts attention in the beauty of its blue. Here, however, it is of the deepest Italian, and the flowers are no larger than those of the forget-me-not. Every part of this curious plant is rough and harsh with hairs strong enough to penetrate the skin.

Very welcome is the contrast supplied in the beautiful golden Chlora, *Chlora perfoliata*; and in its near ally, and frequent companion, the rose-coloured *Erythræa*. These two plants are esteemed in domestic medicine as tonics. In days gone by it was quite a common thing to see operatives from the manufacturing districts who had been awhile at the "Strangers' Charity," busy collecting armfuls of each, to be carefully taken home, and there preserved for use. The Chlora is at once distinguished from everything else upon the sandhills, by the smooth-edged leaves being coupled in such a way as for every pair to seem but *one* leaf, the stem passing through, and a handsome terminal cluster of star-shaped flowers of the purest and clearest golden yellow. The *Erythræa*, commonly called "Sanctuary," a rustic corruption of "centaury," has narrow leaves, also in pairs, but quite free, and the star-shaped flowers, again in terminal clusters, are pink. This very pretty plant has two well-marked varieties, for the possession of which the Southport sandhills were, with botanists, formerly celebrated. The *pulchella* is probably still there; the *latifolia* very doubtfully.

Where large broad plateaux, moist and permanently green, are found among the sandhills, though some of the above-named plants are not excluded, others of quite new character and singularly charming are found in plenty. Foremost among these are the Parnassia and the Pyrola, both of the purest white, and in delicacy unexcelled by any garden flower. The blossoms of the Parnassia *(P. palustris)* are shaped like those of the common buttercup, and borne upon the tips of slender stalks varying from three to six inches in height. They often grow in companies of ten to twenty or more, all rising from the same root, which may easily be dug up for conveyance home, where, if got young, and carefully placed in a flower-pot and kept well watered, the beautiful little thing may be retained as a parlour ornament for many weeks. There are localities among the sandhills beyond Birkdale where, in favourable seasons, so vast is the quantity of the Parnassia that the whiteness of the ground may be compared to that given by daisies to the sward.

The Pyrola *(P. rotundifolia)* has been well styled the lily-of-the-valley of the sandhills. This plant also grows in great profusion, so that handfuls are gathered by visitors without any seeming abatement. The leaves are roundish, and nearly all close upon the surface of the ground. The flower-stems are erect, four or five inches in height, and bear at the summit half-a-dozen very pretty white corollas, the odour evolved from which is so powerful that it can be perceived as one walks along. The botanists find a technical difference between the Birkdale plant and the rotundifolia of other parts of England. The Birkdale form they distinguish as the variety *bractescens* or *maritima*, and excepting upon the sandhills at Lytham, a few miles to the north of Southport upon the opposite side of the estuary of the Ribble, it is understood to be entirely confined to this neighbourhood.

Mingling with these two lovely wild-flowers there may be found, in autumn, plenty of the purple gentian, *Gentiana campestris*, or perhaps *Amarella;* still, more probably, both species, the differences being very slight. In spots where the soil is still more decidedly wet during many months of the year there is rivalry, in regard to abundance, on the part of that very curious orchideous plant, the marsh helleborine, *Epipactis palustris.* Though not so showy as many of its race wild in England, the Epipactis is formed and coloured in a remarkably pretty manner, and in itself quite rewards a journey to the Birkdale sandhills, where it blooms towards the end of July. An almost constant associate of this pretty orchid is one of the genuine orchises, the *O. latifolia*, or purple marsh orchis; the flowers borne in a densely contracted though still hyacinthine, cluster. A third orchideous plant, met with upon the inland side of the sandhills, where there is a fair permanent supply of moisture, is the Tway-blade *(Listera ovata)*, more curious than either of the preceding, the stem bearing only two large oval leaves—whence the name—and at the upper part a score or two of little green blossoms so ridiculously like the human figure that the Listera is often mistaken for the genuine Green-man orchis *(Aceras anthropophora)*, but this one occurs only in the south-eastern counties.

In parts of the sandhills range a trifle more distant, but still within the compass of an agreeable walk, there are tracts such as are congenial to semi-amphibious plants. Here may be gathered the beautiful marsh red-rattle, *Pedicularis palustris;* the silver-tassels, prettiest of the cotton-sedges, *Eriophorum polystachyon;* the white brookweed, *Samolus Valerandi;* the *Viola palustris*, and the *Blysmus compressus*.

More or less generally diffused over the sandhills, the observant eye will catch sight also of the following plants:—

Seaside Catstail-grass *(Phleum arenarium)*.

Spring Hair-grass *(Aira præcox)*.

Seaside Fescue-grass *(Festuca uniglumis)*.

Seaside Wheat-grass *(Triticum junceum)*, a large form, it would appear (changed also in complexion by the conditions of its place of growth), of the common couch-grass *(Triticum repens);* and in that case supplying a beautiful example of the versatility of Nature, especially when with such instruments in hand as are supplied by the margin of the sea.

Sweet Yellow Galium, or "Yellow Bedstraw" *(Galium verum)*, a very delicate and elegant little plant, the dark-green leaves as slender as needles, about an inch in length, and spreading from the stem in little circles of rays; the minute flowers in loose light tufts, golden yellow, conspicuous from their abundance, and honey-scented. The Sweet Yellow Galium is the original "Maiden-hair," the name implying a fancied resemblance to the unsnooded locks of girls in the times when light-hued hair was very specially admired, and given by the poets to their heroines, as in *Chaucer:*—

> Her yellow hair was broidered in a tresse.

Buck's-horn Plantain *(Plantago Coronopus)*.

Bog Pimpernel *(Anagallis tenella)*.

Seaside Convolvulus *(Calystegia Soldanella)*. A very elegant trailing convolvulus, with pale rosy flowers.

Bluebell *(Campanula rotundifolia)*. The true "Bluebell of Scotland," often miscalled the "harebell," which latter name belongs rightfully to the sylvan hyacinth, *Scilla nutans*; the inheritance dating from the time of Shakspere, who never confuses the flowers of different seasons, and associates the scilla with the primrose—

> . . . . . . Thou shalt not lack
> The flower that's like thy face, pale primrose, nor
> The azur'd harebell, like thy veins.

Wild Parsnip *(Pastinaca sativa)*. Easily recognised by its coarse and untidy habit, large leaves, and terminal umbels of yellow flowers.

Evening Primrose *(Œnothera biennis)*. An American plant, in reality, but which becoming accidentally dispersed, through the medium of the abundant seeds, in waste and little-travelled ground, has long since become perfectly naturalized, and is to be met with more or less almost all the way to Liverpool.

Sea-side Catchfly *(Silene maritima)*. A maritime form of the common bladder-campion *(Silene inflata)* of inland districts, told at once by its slender trailing stems, small sea-green leaves growing in pairs, and very handsome round snow-white flowers the size of a shilling.

English Catchfly *(Silene Anglica)*.

Evening Star Catchfly *(Silene noctiflora)*. Near Ainsdale.

Golden Stone-crop *(Sedum acre)*. Though for want of walls and cliffs upon which it can take its favourite form of the epaulette, this beautiful little plant cannot be expected to be seen at Southport in perfection;—the growth upon the ground is free and cheerful, and the brilliancy of the individual flowers is in no degree inferior.

Knotted Spurrey (*Spergula nodosa*). Plentiful, and when the sunshine opens the sensitive milk-white flowers, a very pleasing little gem.

Agrimony (*Agrimonia Eupatorium*). The yellow flowers, scented like apricots, borne in a long and erect spire, which never has an immediately adjacent neighbour.

Sandhills Rose (*Rosa spinosissima*, or *pimpinellifolia*). The most thorny of its beautiful genus; the usual height about twenty inches; the large flowers creamy white.

The Dewberry (*Rubus cæsius*). In many places among the sandhills, especially where fenced as private property, this very interesting species of bramble attains perfection. Trailing upon the ground, the long branches reduce one's pace to the slowest. In late summer they are loaded with the handsome fruit, at once distinguished from blackberries by the great size and the fewness of the component drupeolæ, which are covered, moreover, with a delicate glaucous bloom, instead of being jetty and shining.

Wild Thyme (*Thymus Serpyllum*). Like the Golden Stonecrop, this elegant little plant is unable to present the most beautiful forms it is capable of taking for want of broken rock. The hue and the odour, however, are such as always near the sea.

Yellow Bartsia (*Bartsia viscosa*). Occurs in damp localities upon the inland side of the sandhills, commencing with the neighbourhood of Birkdale.

Teesdalia (*Teesdalia nudicaulis*). In many places, after quitting the town southwards.

Spring Draba (*Draba verna*). Upon dry and grassy slopes, similar to those haunted by the Teesdalia.

Flix-weed (*Sisymbrium Sophia*). At Birkdale, among the sandhills, this plant occurs sometimes in incredible profusion. Distinguished at once from all our other native Cruciferæ by the minuteness of the yellow flowers, and the very light and minutely dissected leaves.

Common Stork's-bill (*Erodium cicutarium*). Abundant upon grassy slopes.

Large Crimson Cranes-bill (*Geranium sanguineum*). Occasionally in similar situations.

Milk-wort (*Polygala vulgaris*). Common, where defended by short turf, and in all its curious diversities of colour—violet-purple, lavender-blue, pink, and creamy-white.

Rest-harrow (*Ononis arvensis*). In cloudy weather scarcely noticeable, but when the sun shines warm and bright, very beautiful in the abundance of its large rosy-pink flowers, fashioned like those of the sweet-pea. Occurs both with spines and without.

Lady's Fingers (*Anthyllis Vulneraria*). Another very striking leguminous plant, the heads of lemon-yellow flowers, which bear a good deal of resemblance to clover, invariably two together, one slightly above the other, and the calyces covered with dense white down.

Bird's-foot (*Ornithopus perpusillus*). The most dainty of the indigenous species of the same order, the little pink-veined flowers requiring a magnifier to be appreciated. Occurs upon grassy slopes, where somewhat dry.

Autumnal Hawkbit (*Apargia autumnalis*). In similar situations.

Sand-hills Carex (*Carex arenaria*). Everywhere, in profusion, spreading underground.

Narrow-leaved Hawk-weed (*Hieracium umbellatum*). Formerly very plentiful upon the sandhills to the north of Southport, and doubtless still holding its old seat in quiet recesses.

Sea-side Thistle (*Carduus tenuiflorus*). Upon the sandhills, both north and south, but scarce, the seeds seeming to be a favourite kind with frugivorous birds.

Milk-thistle (*Carduus Marianus*). Occasionally. In respect of foliage, the most beautiful of the English thistles, every leaf being laced and veined with white, whence the name and the mediæval dedication.

Common Cud-weed (*Filago Germanica*). Upon grassy slopes, where somewhat dry. Rather scarce.

Blue Flea-bane (*Erigeron acris*). In similar habitats, and of the same degree of infrequency.

Heath Groundsel (*Senecio sylvaticus*). Often shelters itself upon the inland side of a bed of mat-grass. A very interesting plant, in figure like the common garden groundsel, but strongly aromatic, and the exterior of the capitules wholly green.

Golden-rod (*Solidago Virg-aurea*). Occasionally in dry and grassy places, where not exposed to the direct sea-breeze.

At the foot of the sandhills, upon the seaward side, where touched at periods, more or less distant, by the salt water, or by the spray of high tides; and upon the broad flats which are covered frequently, grow many other plants of singular interest. Some of these never occur except in habitats such as, for convenience sake, may be called saline. A few of them, strange to say, do quite as well in calcareous inland fields. To the first class may be referred the following—some of them frequent, others, at Southport, very scarce :—

Glass-wort (*Salsola Kali*).
Sea-side Poa (*Poa maritima*).
Procumbent Poa (*Poa procumbens*).
Rottböllia (*Rottböllia incurvata*).
Sea-side Plantain (*Plantago maritima*).
Sea Milk-wort (*Glaux maritima*).
Sea-side Goose-foot (*Suæda maritima*).
Sea-lavender (*Statice Limonium*).
Sea-side Arrow-grass (*Triglochin maritimum*).
Sea-side Sand-wort (*Honckneya peploides*).
Yellow-horned Poppy (*Glaucium luteum*).
Purple Sea-rocket (*Cakile maritima*). A beautiful plant, the flowers resembling those of the garden stock, and often to be found, when the season is mild, in the depth of winter.
Sea-side Wormwood (*Artemisia maritima*).
Sea-side Starwort (*Aster Tripolium*). A very charming autumnal flower; the blossoms, yellow, with lilac rays, resembling those of the Michaelmas daisy, or "Farewell-summer" of the gardens, and produced abundantly upon the flats, where often wetted.
Salicornia (*Salicornia herbacea*). Often supposed to be, and collected for sale under the name of "samphire." Distinguished at once by the total want of leaves, the salicornia consisting of green, cylindrical, fleshy pencils, joined, as it were, end to end.

The plants found upon the Southport shore, which also occur in inland habitats, are preëminently the—

Sea-side Pink, or Thrift (*Armeria maritima*).

Scurvy-grass (*Cochlearia Danica*), and the

Strawberry Trefoil (*Trifolium fragiferum*), that pretty species which converts the heads of bloom during the maturation of the seed into the similitude of raspberries.

THE INLAND PLANTS.

The inland portion of the Southport flora consists, in the main, of plants such as are commonly met with all over England, in fields and by waysides — those, in a word, which constitute the ordinary vegetation of the country, where a distinct character is not given by the predominance of limestone. Plants requiring or fond of calcareous soil are here not to be expected. In compensation there occur many which are identified more particularly with the New red-sandstone formations, and which, in course of ages, have no doubt moved westwards from the great red-sandstone districts a few miles further to the east. To enumerate these very common and universally diffused plants is not necessary. The list would be little more than a catalogue of the ordinary components of turf, and of the accustomed weeds of cultivated land, and of shaded banks and rural hedgerows, though many of the latter are quite as well entitled to the appellation of wild-flowers as the rarest and most eagerly sought of the botanical treasures and curiosities. Many of these comparatively common plants are not, after all, to be expected to occur near Southport in abundance. A walk of two or three miles into the country is needful for the finding of a considerable number. The direction likely to prove most rewarding is that one which would be taken by anyone bent on learning the country about Crossens and Churchtown, and beyond, up to the borders of the Ribble.

The most interesting of the inland plants found where the ground is dry—a class quite distinct from the aquatics and the amphibia—are as follow:—

Germander Speedwell (*Veronica Chamædrys*).
Medicinal Tea Speedwell (*Veronica officinalis*).
Sweet-scented Vernal-grass (*Anthoxanthum odoratum*).
Meadow Fox-tail Grass (*Alopecurus pratensis*).
Meadow Cat's-tail Grass (*Phleum pratense*).
Brown Bent-grass (*Agrostis vulgaris*).
Meadow Soft-grass (*Holcus lanatus*).
Silver Oat-grass (*Arrhenatherum avenaceum*).
Golden Oat-grass (*Trisetum flavescens*).
Smooth Meadow-grass (*Poa pratensis*).
Rough Meadow-grass (*Poa trivialis*).
Quaking-grass (*Briza media*).
Rough Cock's-foot Grass (*Dactylis glomerata*).
Crested Dog's-tail Grass (*Cynosurus cristatus*).
Meadow Fescue-grass (*Festuca pratensis*).
Soft Brome-grass (*Bromus mollis*).
Ray-grass (*Lolium perenne*).
Meadow Burnet (*Sanguisorba officinalis*).
Lady's Mantle (*Alchemilla vulgaris*).
Viper's Bugloss (*Echium vulgare*). The sight of a mass of this plant one summer at Birkdale filled everybody with astonishment.
Scarlet Pimpernel (*Anagallis arvensis*).

Primrose (*Primula vulgaris*).
Cowslip (*Primula veris*).
Black Nightshade (*Solanum nigrum*).
White Hedge Bindweed (*Calystegia sepium*).
Sheep's Scabious (*Jasione montana*).
Woodbine (*Lonicera Periclymenum*).
Dog's Violet (*Viola canina*).
Wild Pansy (*Viola tricolor*).
Wild Carrot (*Daucus Carota*).
Lilac Hedge-parsley (*Torilis Anthriscus*)
Pig-nut (*Bunium flexuosum*).
Slender Hedge-parsley (*Anthriscus vulgaris*).
Bladder Campion (*Silene inflata*).
Stellaria (*Stellaria Holostea*).
Wood-sorrel (*Oxalis Acetosella*).
Corn-cockle (*Githago segetum*).
Red Campion (*Lychnis dioica*).
Wood Strawberry (*Fragaria vesca*).
Silver-weed (*Potentilla Anserina*).
Cinque-foil (*Potentilla reptans*).
Corn-poppy (*Papaver dubium*).
Wood Anemone (*Anemone nemorosa*).
Yellow Toad-flax (*Linaria vulgaris*).
Foxglove (*Digitalis purpurea*).
Lady-smock (*Cardamine pratensis*).
Yellow Rocket (*Barbarea vulgaris*).
Herb-Robert (*Geranium Robertianum*).
Musk Mallow (*Malva moschata*).
Fumitory (*Fumaria officinalis*).
Furze (*Ulex europæus*).

Broom (*Sarothamnus Scoparius*).
Bitter Vetch (*Orobus tuberosus*).
Yellow Field-pea (*Lathyrus pratensis*).
Purple-tufted Vetch (*Vicia Cracca*).
Narrow-leaved Crimson Vetch (*Vicia angustifolia*).
Melilot (*Melilotus officinalis*).
Zigzag Clover *Trifolium medium*).
Meadow Lotus *Lotus corniculatus*). Very beautifully developed; also upon the sandhills.
Common St. John's Wort (*Hypericum perforatum*).
Pretty St. John's Wort (*Hypericum pulchrum*).
Yellow Goat's-beard (*Tragopogon pratensis*).
Mouse-ear Hawkweed (*Hieracium Pilosella*).
Golden-rod (*Solidago Virgaurea*).
Flea-bane (*Pulicaria dysenterica*).
Ox-eye Daisy (*Leucanthemum vulgare*).
Corn-marigold (*Chrysanthemum segetum*).
Sneeze-wort (*Achillea Ptarmica*).
Milfoil (*Achillea Millefolium*).
Common Centaury (*Centaurea nigra*).
Wild Hop (*Humulus Lupulus*).
Tamus (*Tamus communis*).

The level land which stretches east and southwards from Southport is in many parts intersected by little watercourses, the flora of which presents features quite novel, and often of great attractiveness. The most interesting constituents are not, perhaps, those which possess the special charm of beauty of blossom, except in some few instances. There are plenty, however, that are gay enough for the bouquet, as will be seen from the enumeration, which includes also the principal marsh plants.

Mare's-tail (*Hippuris vulgaris*). This very curious plant occurs in ditches near Blowick, and in various other places; also in the little pools retained in gardens. Those who are interested in productions so singular should be careful not to overlook the submerged shoots, developed in their perfection about October, when the leaves are several inches in length.

Marsh Speedwell (*Veronica scutellata*).

Water Speedwell (*Veronica Anagallis*).

Brooklime (*Veronica Becabunga*).

Bladderwort (*Utricularia minor*).

Ivy-leaved Duckweed (*Lemna trisulca*).

Great Duckweed (*Lemna polyrhiza*).

Great Lilac Valerian (*Valeriana officinalis*).

Yellow Water-flag, or Fleur-de-lis (*Iris Pseud-acorus*).

Ribbon-grass (*Digraphis arundinacea*).

Tufted Hair-grass (*Aira cæspitosa*).

Floating Sweet-grass (*Glyceria fluitans*).

Great English Reed (*Phragmites communis*).

White Marsh Galium (*Galium palustre*)

Pondweed (*Potamogeton*). Several species occur, but they wait discrimination.

Poets' Forget-me-not (*Myosotis palustris*).

Hottonia (*Hottonia palustris*). One of the most beautiful of our native aquatics, the slender flower stems rising to a height of about nine inches above the surface of the water, with many whorls of flowers resembling lilac primroses.

Bogbean (*Menyanthes trifoliata*). A worthy companion of the preceding, and its near rival in loveliness, as may be judged from one of its names, "the hyacinth of the marshes." Distinguished at once by the white beards upon the purple-tinged white petals.

Marsh Violet (*Viola palustris*).
Wild Celery (*Apium graveolens*).
Angelica (*Angelica sylvestris*).
Slender Water Dropwort (*Œnanthe fistulosa*). A very singular and elegant plant, composed in every part of slender green tubes.
Common Water Dropwort (*Œnanthe crocata*).
Common Marshwort (*Sium nodiflorum*).
Fine-leaved Marshwort (*Sium angustifolium*).
Common Water-Plantain (*Alisma Plantago*).
Small Water-Plantain (*Alisma ranunculoides*).
Amphibious Persicaria (*Polygonum amphibium*).
Flowering-rush (*Butomus umbellatus*).
Ragged Robin (*Lychnis Flos-cuculi*).
Purple Lythrum (*Lythrum Salicaria*).
Meadow-sweet (*Spiræa Ulmaria*).
Water Septfoil (*Comarum palustre*).
White Water-lily (*Nymphæa alba*).
Yellow Water-lily (*Nuphar lutea*).
Water Snowcups (*Ranunculus aquatilis*).
Marsh-marigold (*Caltha palustris*).
Common Water-mint (*Mentha hirsuta*). Occurs also among the sandhills.
Water-cress (*Nasturtium officinale*).
Bur-marigold (*Bidens cernua*).
Butterbur (*Petasites vulgaris*).
Reed-mace (*Typha latifolia*).

Common Bur-reed (*Sparganium ramosum*).

Sedges (*Carex*). The species found wild about Southport wait discrimination.

Common Horn-wort (*Ceratophyllum*).

Water Milfoil (*Myriophyllum*).

Cinnamon-rush (*Acorus Calamus*).

Frog-bit (*Hydrocharis Morsus-ranæ*).

FERNS AND FERN-ALLIES.

The neighbourhood of Southport, it hardly needs the saying, is not one which can be expected to be favourable to an abundance of native ferns. Ferns are plants which in most cases love moist and shady woods, or calcareous cliffs, or, as happens with the sub-maritime species, rocks like those of North Devon. Habitats such as these are not compatible with the sandy flats of the sea-margin of Lancashire, though at a few miles distance inland, it is not difficult to discover groves where the larger kinds of ferns, at least, have a congenial abode. The beautiful, not to say romantic dell, or ravine, at Gathurst, called Dean-wood, teems with the graceful forms of the shield-ferns, in several kinds. The deep cuttings upon the borders of the moss, especially towards Halsall and Scarisbrick, likewise afford a plentiful display; while towards Crossens, and thereabouts, the hedgebanks also contribute. The following are the species which have been observed:—

Common Polypody (*Polypodium vulgare*).

Common Shield-fern (*Lastrea Filix-mas*).

Broad-leaved Sylvan Shield-fern (*Lastrea dilatata*). The most plentiful of the South and West Lancashire ferns.

Lady-fern (*Asplenium Filix-fœmina*).
Common Brake (*Pteris Aquilina*).
Northern Hard-fern (*Blechnum boreale*)..
Osmunda (*Osmunda regalis*).
Moonwort (*Botrychium Lunaria*).
Adder's-tongue (*Ophioglossum vulgatum*).

EQUISETACEÆ.

Three or four species of the very curious genus Equisetum occur about Southport, especially at and beyond Birkdale and Ainsdale, where, in the slacks among the sandhills, there is great plenty, in particular, of the *Equisetum palustre*. The very pretty little *variegatum* likewise occurs here. In cultivated land, and in gardens as a troublesome weed, the *arvense* is also well known; and in ponds may be found, at intervals, the densely-crowded and reed-like *limosum*.

LYCOPODIACEÆ.

One only of this remarkable order belongs to the Southport flora, the very elegant little *Lycopodium selaginoides*, which is met with occasionally upon the sandhills. The habitats of the other British species are such as to render their existence at Southport impossible.

MOSSES.

To the bryologist, as said above, Southport has many attractions, though limited, as regards locality, to the sandhills. So much remains still to be done in the examination of our local mosses, that no more than a very partial list could at present be given. We consider it better, accordingly, to wait until some painstaking student shall supply a catalogue that will be at least approximately complete.

The same may be said as regards the local Algæ, Lichens, Fungi, and the other families of the purely cellular cryptogamia. Southport is still in expectancy that her scientific men will devote a portion of their time to these most beautiful of the minor forms of vegetable life. A rich harvest awaits their researches. Everyone who possesses a good microscope may help in the good work.

CHAPTER VI.

Ten thousand warblers cheer the day, and one
The live-long night; not these alone, whose notes
Nice-fingered art must emulate in vain,
But cawing Rooks, and Kites that swim sublime
In still repeated circles screaming loud:
The Jay, the Pie, and e'en the boding Owl
That hails the rising moon, have charms for me.

COWPER.

THE SOUTHPORT BIRDS.

THE interest of the Southport Ornithology equals that of its Botany. That many of the most celebrated of our native British birds are never seen or heard in the neighbourhood, is no doubt true. Those which belong to us, it is equally true, are the ordinary ones of our sea-girt island. To anyone who really cares for birds, this makes no difference. Their being ancient Britons in no degree diminishes the interest attaching to the observation of their habits, manners, and customs; nor, if they be of the minstrel class, does it render their song less inviting and reanimating. Should the nightingale some day pay us a visit, the fact would be a very interesting one to record in the newspaper; meanwhile, it is incomparably better that we have the permanent and substantial reality of the presence of representatives of the best of the genuine old "county families"—the throstle, the little dunnock, and the skylark. On a summer's morning, where shall we look for sensations more delightful than are excited by the little creature that "at heaven's gate sings?"

> A charm from the skies seems to hallow us there,
> Which, seek through the world, you'll not meet with elsewhere.

Many very interesting birds come as visitors also, in spring and summer—the swallow, the whitethroat, the yellow wagtail, the wheatear; and, best of all—because fitted by nature to give pleasure to the largest number of people—the cuckoo, sweet magnet of the heart, as in May we tread the rising grass, or wander amid the airy solitudes of the sandhills, which provide enjoyment for all seasons, and in no way more bountifully than through the medium of the birds. In winter, again, there is an influx of visitors by no means inconsiderable, these consisting chiefly of curious and uncommon shore-birds, the variety of which belonging to the northern coast of Lancashire, gives it very special attractions to the ornithologist. Exposed to wind and sea-tempest, the shore, in winter, often furnishes other very interesting specimens—the bodies of maritime birds that have succumbed to distress of weather, and are tossed up like so much wreckage. The puffin, the razor-bill, and the stormy petrel, in particular, are thus made known to observers intent on such things. Curious sea-birds also get captured sometimes by fishermen, when far out upon the water, not uncommonly by entanglement in their nets; and these, when of eatable kinds—being chiefly young birds—get exposed for sale in the market.

The number of distinct species of birds known to exist in the world is not less than eleven thousand, and many more probably exist in remote corners, of which little has yet been learned. In Great Britain, there exist about two hundred, either established denizens, or regular visitors; and about a hundred and sixty more have been known to come for awhile, and at longer or shorter intervals, from other countries—some, as rare emigrants, others, it would seem, by misadventure. Birds, in their aërial voyages, often wander inconceivably far from home; so that, in all countries, solitary examples of different kinds are met with in turn, and this, once for all, or nearly so—the same spot being never revisited. In 1807, a pratincole came to Ormskirk, where it was shot, as a reward for its love of adventure, the stuffed remains going to Knowsley, and thence to the Liverpool Museum. A red phalarope came to Birkdale in 1832. Since about that time, the ring-ousel and the arctic gull have also come to Southport, once each.

In surveying the ornithology of any given district, it becomes important accordingly to distinguish the birds of the year into three separate classes,—Permanent Residents (in Britain); the Regular Visitors from other countries, summer or winter; and the Casuals, or vagrants. Of the first class, Southport appears to have about sixty; regularly immigrating birds have been reckoned to the number of about twenty; and of Casuals, chiefly shore and web-footed birds, the list runs to about fifty. Many of the kinds of birds, however, which no further back than twenty-five years ago, were plentiful in the neighbourhood of Southport, have shared the fate of the plants once so abundant, which now are scarce. Much waste and marshy land has been drained, and adapted to the requirements of a constantly increasing population; and the birds, finding their breeding places destroyed, and their feeding localities impoverished, have wisely betaken themselves elsewhere. There has also been much deliberate destruction with the gun—a form of persecution very often unnecessary and cruel. Farmers and gamekeepers are now happily becoming alive to the fact that birds such as kestrels and owls are not only harmless, but useful; their food consisting chiefly of mice and other ground vermin. Whereever hawks have been ruthlessly shot down, to the almost total extermination of their kind, agriculturists have sustained heavy loss. So with the reckless slaughter of the smaller descriptions of birds. The havoc made in orchards and gardens, and upon farm land, is sometimes undeniably vexatious. But it is balanced by the enormous consumption of insects, grubs and caterpillars. Look, also, at those heaps of broken and excavated snail-shells upon the sand-hills, the picked bones of their repast, left by the thrushes and the blackbirds; both of which, we may remember again, if disposed to be very cross with them, about the nibbling of the cherries, are liable to be themselves eaten by their own pursuers, and very often do get eaten. No policy is more short-sighted than the persistent destruction of birds. It is the opprobrium of the present day, and if not changed, will induce results that, when too late, will be deplored. Severe winters, not so much through the intensity of the cold, as the diminution of the natural food supply, keep all in order, taking the average of years, in regard to excessive increase.

The establishment of what will before long become fine arboretums—the Hesketh Park, and the Churchtown Botanic Gardens—will do much for our local feathered friends. There, at least, they will be safe; a good effort of one kind undesignedly promoting advantages of another and not less desirable order. The maritime birds, fortunately, are not exposed to the same kind of interference as those living on land. The pearly-bosomed sea-gulls have it all their own way; and, here, it may be mentioned that one of the most interesting spectacles to be witnessed in Southport, is the daily feeding of these pretty, though harsh-voiced, birds at the end of the Pier. That, while the land-birds are foremost among nature's sanitary police, in regard to insects, &c., the sea-gulls are the ocean-scavengers, hardly needs the saying. The practice at Southport is to feed the gulls with the refuse of the market fish. Every day at noon, except upon Sundays, it is thrown upon the water, piece after piece, from the lower portion of the stage. The gulls are already assembled—they know it is twelve o'clock; as an example of memory in birds, a better and more interesting proof cannot be found—as regularly as the hour comes, if the day be bright, flocks of them may be distinguished sailing up from a league distance, and for ten or fifteen minutes, the clamour, and then the activity, is an event never to be forgotten.

For the enjoyment to the full, of the song of the skylark, it is best, perhaps, to go to Crossens, and beyond; where the fields and the heavens, at times, are quite saturated with the simple music. A capital collection of specimens of the Southport birds, formed by Mr. George Davis, is contained in the Museum of the Churchtown Gardens.

The following list of the Southport birds includes all the names mentioned in the original edition of this little volume. They were inserted, almost wholly, upon the authority of two excellent practical men, the late Mr. Graves, and the late Mr. Tyrer. The frequency of the occurrence, at the present day, of various species, will, no doubt, be found less by new observers than it was twenty years ago. It shows, in any case, that the time was when Southport could lay claim to knowledge of no fewer than a hundred and thirty representatives of the ornithological department of nature ; and, though they may now be rare, what kinds of birds may reasonably be expected to show themselves.

In this new catalogue, the names are re-arranged into two out of the three great classes above indicated, viz., Birds permanently residing in Britain, and Visitors, coming regularly from foreign countries. The Southport "Casuals" (all accounted British) are marked with an *

To facilitate the studies of persons interested in ornithology, reference is made after every name, to the coloured drawing in Rev. F. O. Morris's " British Birds," six volumes, 8vo, the first dated 1863. The numbers of the plates, as here given, presume them to be consecutive from plate 1 in vol. 1, to plate 358 in vol. 6.

BIRDS PERMANENTLY RESIDING IN BRITAIN.

The Kestrel (*Falco Tinnunculus*).—Morris, vol. 1, pl. 17. Abundant upon the sandhills, where it may be readily distinguished, when upon the wing, by hovering over its prey.

* The Merlin (*Falco Æsalon*).—Vol. 1, pl. 16.

The Sparrow-hawk (*Accipiter fringillarius*).—Vol. 1, pl. 19.

Occasionally upon the sandhills, where it feeds, like the kestrel, upon small birds and young rabbits.

The Short-eared Owl (*Strix brachyotus*).—Vol. 1, pl. 23. Inland, upon the moss, and thereabouts. An autumnal and winter bird, seen upon dark days, hunting the ground in search of mice, shrews, and small birds.

The Brown or Screech Owl (*Strix aluco*).—Vol. 1, pl. 28. Occasionally among the sandhills.

The Barn or White Owl (*Strix flammea*).—Vol. 1, pl. 29. Often to be seen on fine moonlight nights, hunting like the short-eared owl.

The Song Thrush (*Turdus musicus*).—Vol. 3, pl. 127.

The Missel Thrush, Storm Thrush or Rain-bird (*Turdus viscivorus*).—Vol. 3, pl. 124. Acquired its name from its habit of singing during storms. Feeds upon ivy berries, snails, slugs, &c.

The Blackbird (*Turdus merula*).—Vol. 3, pl. 131. Occasionally in the valleys among the sandhills. Fond of plantations and orchards.

The Hedge Sparrow or Dunnock (*Accentor modularis*).—Vol. 3, pl. 135.

The Robin (*Sylvia rubecula*).—Vol. 3, pl. 136.

* The Stonechat (*Sylvia rubicola*).—Vol. 3, pl. 140. Occasionally about Birkdale and Ainsdale, frequenting the tops of furze and other bushes.

The Great Titmouse or Tomtit (*Parus major*).—Vol. 1, pl. 36.

The Blue Titmouse (*Parus cœruleus*).—Vol. 1, pl. 39.

The Marsh Titmouse (*Parus palustris*).—Vol. 1, pl. 40. Haunts the willow and poplar trees.

The Long-tailed Titmouse (*Parus caudatus*).—Vol. 1, pl. 41.

The Water Wagtail, Pied Wagtail, or Dish-washer (*Motacilla Yarrellii*).—Vol. 2, pl. 80.

The Grey Wagtail (*Motacilla sulphurea*).—Vol. 2, pl. 82. Generally in small flocks, and without shyness.

The Meadow Pipit or Titling (*Anthus pratensis*).—Vol. 2, pl. 86. Fond of grassy spots among the sandhills. Also in meadows.

The Skylark or Lavrock (*Alauda arvensis*).—Vol. 2, pl. 93.

The Common Bunting (*Emberiza miliaria*).—Vol. 2, pl. 97.

The Black-headed Bunting (*Emberiza Schœniculus*).—Vol. 2, pl. 98.

The Yellow Ammer, or Yellow Bunting (*Emberiza citrinella*) Vol. 2, pl. 98.

The Chaffinch or Spink (*Fringilla cœlebs*).—Vol. 2, pl. 102.

The House Sparrow (*Passer domesticus*).—Vol. 2, pl. 105.

The Greenfinch (*Coccothraustes chloris*).—Vol. 2, pl. 106.

The Common or Brown Linnet (*Fringilla cannabina*).—Vol. 2, pl. 110.

The Less Redpole (*Linaria minor*).—Vol. 2, pl. 111.

The Starling or Shepster (*Sturnus vulgaris*).—Vol. 3, pl. 121.

The Hooded Crow (*Corvus cornix*).—Vol. 1, pl. 53.

The Rook (*Corvus frugilegus*).—Vol. 1. pl. 54.

* The Jackdaw (*Corvus monedula*).—Vol. 1, pl. 55. Occasionally seen about Halsall and Scarisbrick, in company with rooks and gulls; and about Formby, associating with gulls and terns.

* The Magpie (*Pica caudata*).—Vol. 1, pl. 56.

The Common Wren (*Sylvia Troglodytes*).—Vol. 3, pl. 160.

The Peewit or Lapwing (*Vanellus cristatus*).—Vol. 4, pl. 192.

The Ring-Dove, Cushat, or Wood-Pigeon (*Columba palumbus*).—Vol. 3, pl. 164. Occasionally in considerable numbers in ploughed fields about Scarisbrick and Halsall, consorting with rooks, jackdaws, gulls, and other birds; during winter often seen in turnip-fields.

The Common Partridge (*Perdrix cinerea*).—Vol. 3, pl. 174.

* The Greater Butcher Bird (*Lanius excubitor*).—Vol. 1, pl. 33. Occurs upon the sandhills, apparently searching for lizards, which, when obtained, it transfixes upon thorns, and tears to pieces. A very fierce bird, in spring the terror of all its smaller neighbours, and pursuing any one of them that may approach its place of resort.

The Smaller Butcher Bird or Red-backed Shrike (*Lanius collurio*).—Vol. 1, pl. 34. Like the preceding, this bird impales its prey upon thorns, using the spines of dead thistles, when it captures insects.

* The Hoopoe (*Upupa epops*).—Vol. 1, pl. 49. Once at Birkdale.

* The Rose-Ouzel (*Pastor roseus*).—Vol. 2, pl. 120. "Near Ormskirk."

* The Snow Bunting (*Plectophanes nivalis*).—Vol. 2, pl. 95. Occasionally, in very severe weather, upon farm-land, and among the reeds in the slacks between Ainsdale and Formby.

* The Goldfinch (*Fringilla carduelis*).—Vol. 2, pl. 108. Has been seen upon the Birkdale sandhills, feeding on the seeds of the Carline thistle.

The Golden-crested Wren (*Regulus cristatus*).—Vol. 3, pl. 162. Towards Churchtown.

The Swift (*Hirundo apus*).—Vol. 2, pl. 73.

H

The Heron (*Ardea cinerea*).—Vol. 4, pl. 197. "The larches and birches at Scarisbrick Hall contain about two dozen of the nests."—*Manchester Guardian*, December 28, 1881.

The Bittern (*Botaurus stellaris*).—Vol. 4, pl. 204. More frequently heard than seen, the sound produced by this bird resembling a heavy note upon a drum.

The Curlew (*Numenius arquata*).—Vol. 4, pl. 211. Frequent upon the shore, especially near the Ribble and the Alt. In autumn, often met with in stubble-fields, searching for snails, worms, slugs, and scattered grain.

The Whimbrel (*Numenius phæopus*).—Vol. 4, pl. 212.

The Redshank (*Scolopax calidris*).—Vol. 4, pl. 214.

The Godwit (*Scolopax lapponica*).—Vol. 4, pl. 222.

The Dunlin or Sea-lark (*Tringa alpina*).—Vol. 4, pl. 240.

The Little Stint (*Tringa minuta*).—Vol. 4, pl. 236.

The Knot (*Tringa canutus*).—Vol. 4, pl. 232.

The Turnstone (*Tringa interpres*).—Vol. 4, pl. 193.

The Ruff (*Tringa pugnax*).—Vol. 4, pl. 224.

The Grey Plover (*Tringa squatarola*).—Vol. 4, pl. 191. In the winter not uncommon.

The Golden Plover (*Charadrius pluvialis*).—Vol. 4, pl. 186. Occurs in the slacks about Ainsdale, summer and winter.

The Ring Plover (*Charadrius Hiaticula*).—Vol. 4, pl. 188. Upon the sandhills at Birkdale and Ainsdale.

The Sanderling (*Calidris arenaria*).—Vol. 4, pl. 194.

* The Curlew-billed Sandpiper (*Scolopax pygmæa*).—Vol. 4, pl. 231. Occasionally in the autumn.

* The Water Rail (*Rallus aquaticus*).—Vol. 5, pl. 246.

The Oyster Catcher (*Hæmatopus ostralegus*).—Vol. 4, pl. 195. Upon the shore during the winter.

* The Coot (*Fulica atra*).—Vol. 5, pl. 248.

* The Common Gallinule or Water Hen (*Gallinula chloropus*).—Vol. 5, pl. 247.

The Spotted Water Hen (*Gallinula Porzana*).—Vol. 5, pl. 243.

* The Crested Grebe (*Colymbus cristatus*).—Vol. 5, pl. 294. Formby.

* The Eared Grebe (*Colymbus auritus*).—Vol. 5, pl. 297.

* The Little Grebe or Dabchick (*Colymbus Hebridicus*).—Vol. 5, pl. 298. Crossens.

* The Avocet (*Recurvirostra Avocetta*).—Vol. 4, pl. 220.

* The Little Auk (*Uria minor*).—Vol. 6, pl. 306.

The Foolish Guillemot (*Uria Triole*).—Vol. 6, pl. 302. Occasionally seen at sea, in flocks of four or six. They dive so rapidly that to obtain specimens is difficult. Sometimes found entangled in fishing nets.

* The Common Cormorant (*Pelicanus Carbo*).—Vol. 6, pl. 310. Has been seen at Formby.

The Gannet or Solan Goose (*Sula Bassana*).—Vol. 6, pl. 312. Frequently seen off the coast in winter.

* The Pochard or Red-headed Widgeon (*Anas ferina*).—Vol. 5, pl. 282.

* The Shieldrake or Sheldrake (*Anas Tadorna*).—Vol. 5, pl. 266. Has been known to breed in the Ainsdale sandhills.

* The Shoveller Duck (*Anas clypeata*).

The Wild Duck or Mallard (*Anas Boschas*).—Vol. 5, pl. 270.

The Widgeon (*Anas Penelope*).—Vol. 5, pl. 273.

The Teal (*Anas crecca*).—Vol. 5, pl. 272.

The Wild or Grey Goose (*Anas anser*).—Vol. 5, pl. 251.

The Greater Black-backed Gull (*Larus marinus*).—Vol. 6, pl. 337.

The Less Black-backed Gull (*Larus fuscus*).—Vol. 6, pl. 336.

The Herring Gull (*Larus argentatus*).—Vol. 6, pl. 338.

The Common Gull or Sea-mew (*Larus canus*).—Vol. 6, pl. 334. Often resorts, as the other gulls sometimes do, to fields and ploughed land, in search of food.

The Kittiwake (*Larus Rissa*).—Vol. 6, pl. 340.

The Black-headed Gull (*Larus ridibundus*).—Vol. 6, pl. 331. Both on the shore and inland, very numerous.

* The Arctic Gull (*Larus parasiticus*).

The Roseate Tern (*Sterna Dougallii*).—Vol. 6, pl. 315. Often seen among the sandhills at Ainsdale and Formby, where it breeds.

The Common Tern or Sea Swallow (*Sterna Hirundo*).—Vol. 6, pl. 316. With the preceding.

* The Black Tern (*Sterna nigra*).—Vol. 6, pl. 323. With the preceding.

* The Little Tern (*Sterna minuta*).—Vol. 6, pl. 322. Occasionally in the same localities.

PERIODICAL VISITORS TO BRITAIN FROM OTHER COUNTRIES, IN SPRING AND SUMMER.

The Wheat-ear (*Sylvia Œnanthe*).—Vol. 3, pl. 142. Among the sandhills in March and April, and again in September and October. This bird is apt to turn over the refuse left by the tide, in search of food.

The Sedge Warbler or Reed Wren (*Sylvia Salicaria*).—Vol. 3, pl. 145.

The Black-cap Warbler (*Sylvia atricapilla*).—Vol. 3, pl. 150.

The Common White-throat (*Sylvia cinerea*). — Vol. 3, pl. 153.

The Whinchat (*Sylvia rubetra*).—Vol. 3, pl. 141. Abundant in the meadows.

The Willow Warbler or Yellow Willow Wren (*Sylvia Trochilus*).—Vol. 3, pl. 156.

The Chiff-chaff or Less Petty-chaps (*Sylvia rufa*).— Vol. 3, pl. 158.

The Yellow Wagtail (*Motacilla flava*).—Vol. 2, pl. 84.

The Cuckoo (*Cuculus canorus*).—Vol. 2, pl. 71.

The Swallow (*Hirundo rustica*).—Vol. 2, pl. 76.

The Sand Martin (*Hirundo riparia*).—Vol 2, pl. 79.

The House Martin (*Hirundo urbica*).—Vol. 2, pl. 78.

The Dotterel (*Charadrius morinellus*).—Vol. 4, pl. 187.

The Spotted Fly-catcher (*Muscicapa grisola*).—Vol. 1, pl. 44.

The Common Sandpiper (*Tringa Hypoleucos*).—Vol. 4, pl. 217.

The Corncrake or Landrail (*Crex pratensis*).—Vol. 5, pl. 242

* The Quail (*Perdrix coturnix*).—Vol. 3, pl. 178.

* The Red Phalarope (*Phalaropus hyperboreus*).—Vol. 5, pl. 250.

* The Grey Phalarope (*Phalaropus lobatus*).—Vol. 5, pl. 249.

IN AUTUMN AND WINTER.

The Fieldfare (*Turdus pilaris*).—Vol. 3, pl. 125.

The Redwing (*Turdus iliacus*).—Vol. 3, pl. 126.

The Common Snipe (*Scolopax gallinago*).—Vol. 4, pl. 227.

The Jacksnipe or Judcock (*Scolopax gallinula*).—Vol. 4, pl. 228.

The Woodcock (*Scolopax rusticola*).—Vol. 4, pl. 225.

The Scoter or Black Douker (*Anas nigra*).—Vol. 5, pl. 279. Often at sea.

The Golden-eyed Duck (*Anas clangula*).—Vol. 5, pl. 288.

* The Scaup Duck (*Anas marita*).—Vol. 5, pl. 254.

* The Tufted Duck (*Anas fuligula*).—Vol. 5, pl. 285.

The Pintail Duck (*Anas acuta*).—Vol. 5, pl. 269.

* The Wild Swan (*Anas cygnus*).—Vol. 5, pl. 261. In very severe weather.

The Barnacle Goose (*Anas Bernicla*).—Vol. 5, pl. 255.

The Brent Goose (*Anas Brenta*).—Vol. 5, pl. 256.

The Goosander (*Mergus Merganser*).—Vol. 5, pl. 293.

* The Red-breasted Goosander (*Mergus serrator*).—Vol. 5, pl. 292.

The Smew (*Mergus albellus*).—Vol. 5, pl. 290.

* The Northern Diver (*Colymbus glacialis*).—Vol. 6, pl. 299.

The Red-throated Diver (*Colymbus septentrionalis*).—Vol. 6, pl. 301. Not uncommon off the coast, and occasionally taken in nets.

* The Sandwich Tern (*Sterna Boysii*).—Vol. 6, pl. 314.

CHAPTER VII.

It was the blush of morn, earth's choral hour,
And the green grass was veil'd with gossamer,
Silken as faëry tunics seen in dreams,
And set with dew-pearls, fairer far than ours!
What loom can emulate the spider's craft,
Or weave, as they have woven thus, all time?
We call them loathsome, cruel—who can look
Upon the jewell'd Diadema, thron'd
Within her complex armature of toils,
And fail to wonder? Who hath arm'd this race
With all the lithesome serpent's fatal craft?
Set them by glebe and woodland, pool and cave
The ancient, peerless hunters of the world?

B. CARRINGTON.

ARACHNIDA AND CRUSTACEA OF SOUTHPORT.

THE entire credit of the list of local spiders here given is due to the Rev. O. Pickard-Cambridge, one of the greatest of the British authorities upon the subject.

The following remarks by Mr. Cambridge are valuable, as being explanatory of the principles on which the list has been compiled:—"I do not pretend to say that this is a perfect list, for on one side of Southport lies a vast tract of fen or moss land, which I have hardly ever had time to search at all; but the ground I have searched, principally the sandhills along the coast, has been ransacked pretty thoroughly; and, therefore, as the area is so much the more confined, the list is perhaps of so much the greater value. The relative abundance of species in any locality is also, I think, of importance; but the words we commonly use to denote abundance or the contrary are generally so vague, and used or understood by different naturalists in so different a sense, that I will just in a few words try to explain the value of the general terms 'rare,' 'common,' etc., appended to the names in the list, as I myself use and understand them.

"The term *very common* is used to denote that the species may be taken, in its season, in the locality in question, as we should say in popular language, 'in any numbers,' that is, that a hundred or so might be captured during an afternoon of four or five hours, and this without any special search for it.

"*Common* denotes that, in popular language, 'a great many' might be taken in the above time, that is, to the number of, say, forty or fifty, and this with but slight special search.

"*Frequent* denotes that a score or so might be taken, in the same time, with ordinary careful search.

"*Not rare* denotes that a close search will generally procure what we call 'a few,' that is, from five to ten or a dozen.

"*Occasional* denotes that during the time stated, and with careful search, two or three may be captured.

"*Rare* would show that a specimen would be likely to be obtained as we should say only 'once now and then,' that is, about once out of several afternoons' very careful search.

"*Very rare* would denote that one or two specimens in the run of a season would be all that a careful search and open-eye for it would obtain."

To assist those who may be disposed to collect in this branch of Natural History, it may be well to state the mode of preserving Spiders. Specimens should be put up in small glass tubes filled with spirits of wine, or what is better still, in small bottles having a slight constriction or neck near the mouth, so that the cork can be compressed and the rapid evaporation of the spirit be prevented.

In the following list it will be seen that of the two tribes of the order Araneidea known to inhabit Great Britain, one only is represented; of the families making up this tribe nine out of ten are represented (the tenth, however, contains but one British genus and one British species); out of twenty-eight genera composing the families eighteen are represented; and lastly, out of two hundred and seventy species contained in the twenty-eight genera, eighty are represented.

FAMILY LYCOSIDÆ.

GENUS LYCOSA.

Agretyca. Frequent; among grass and herbage on banks and sides of ditches, etc.

Campestris. Not rare; in same places as the last.

Andrenivora. Very rare; on sandhills.

Nivalis. Common; on sandhills.

Rapax. Frequent; in company with *Agretyca*.

Pieta. Not rare; on sandhills.

Saccata. Frequent; on moss land, etc., among grass.
Obscura. Occasional; in company with the last.
Exigua. Very common; almost everywhere.
Cambrica. Not rare; among grass in the slacks, but yet very local.
Piratica. Frequent; in same localities as *Cambrica*.

FAMILY SALTICIDÆ.

GENUS SALTICUS.

Scenicus. Not rare; on walls, posts, and palings in sunshine.
Sparsus. Rare; on trees, among grass, stems, and on walls.
Floricola. Very rare; at grass roots on north sandhills.
Frontalis. Frequent; at roots of grass and rubbish, on bank sides.
Cupreus. Very rare; in company with *Frontalis*.
Blackwallii. Very rare; a single adult female of this large handsome species was captured on a gate close to the shore on the south side of the town, by the Rev. Hamlet Clarke, in September, 1855.

FAMILY THOMISIDÆ.

GENUS THOMISUS.

Cristatus. Occasional; on the ground and at grass roots.
Audax. Very rare; on the ground and at grass roots.

GENUS PHILODROMUS.

Cæspiticola. Frequent; on dwarf willows on sandhills.
Oblongus. Common; at roots and on stems of star-grass, etc.

FAMILY DRASSIDÆ.

GENUS DRASSUS.

Pumilus.   Rare ; on bare sandhills and at roots of grass.

Clavator.   Very rare ; under ledges of sandhills and under stones.

Cupreus.   Frequent ; at roots of grass and moss.

Nitens.   Not rare; among rubbish on dry bank sides, etc., the adult males running on roads, etc., in spring.

GENUS CLUBIONA.

Holosericea.   Occasional ; in angles of summer-houses and in curled leaves, etc.

Amarantha.   Frequent ; at roots of star-grass and in curled leaves.

Epimelas.   Rare ; in curled leaves and holes in posts, etc.

GENUS ARGYRONETA.

Aquatica.   In dykes, among water-weed and rubbish.

FAMILY CINIFLONIDÆ.

GENUS CINIFLO.

Atrox.   Not rare ; under ledges of sandhills overgrown with dwarf willow.

Similis.   In outhouses, etc., not rare ; very closely allied to *Atrox*. This is one of our common house spiders.

GENUS ERGATIS.

Benigna.   Very rare ; at tips of shoots of plants, etc., in a web, and running on paths in spring.

Latens.   Rare ; running on ground in spring.

FAMILY AGELENIDÆ.

GENUS AGELENA.

Labyrinthica. Very common; sitting in a tube in the centre of a wide-spread net: all over the willow-grown sandhills.

Brunnea. Not rare; at roots of star-grass and weeds, etc.

Civilis. Frequent; in outhouses and old buildings. This and *Ciniflo similis* are our two common house spiders.

FAMILY THERIDIIDÆ.

GENUS THERIDION.

Lineatum. Common; almost everywhere.

Quadripunctatum. Rare; in summer-houses and unused rooms.

Nervosum. Not rare; on bushes, etc., in a web.

Pictum. Not rare; on hollies and in greenhouses.

Varians. Frequent; in company with the two last.

Carolinum. Common, though local; in many spots among dwarf willows and herbage on the sandhills.

Pallens. Rare; upon Scotch firs on the moss, at Kirkby.

Variegatum. Occasional; among grass and weeds on dry bank sides, near Churchtown, with its beautiful and pear-shaped nest.

Filipes. Rare; beneath sea-weed on shore.

FAMILY LINYPHIIDÆ.

GENUS LINYPHIA.

Montana. Frequent; on Scotch firs, etc.

Marginata. Frequent; in hedges and in angles of outhouses.

Pratensis. Frequent; on low plants in woods, etc.
Fuliginea. Rare; among star-grass, etc., on sandhills.
Minuta. Not rare; among star-grass, etc., on sandhills, and in porches and unused rooms, etc.
Alticeps. Frequent; among star-grass, etc., on sandhills.
Tenuis. Common; among star-grass, etc., on sandhills.
Terricola. Common; among star-grass, etc., on sandhills: very closely allied to *tenuis*.
Anthracina. Rare; among star-grass, etc., on sandhills.
Pulla. Rare; among star-grass, etc., on sandhills.
Ericœa. Frequent; among star-grass, etc., on sandhills.
Tenella. Very rare; among star-grass, etc., on sandhills.

GENUS NERIENE.

Bicolor. Frequent; at roots of star-grass on sandhills.
Gracilis. Occasional; running on walks, rails, and pavements.
Cornuta. Occasional; among grass, etc., on sandhills.
Apicata. Very rare; among grass, etc., on sandhills.
Longipalpis. Common; among grass, under sea-weed, and on pavements.
Fusca. Rare; under sea-weed in autumn.
Agrestis. Rare; under sea-weed in autumn.
Vigilax. Very rare; among grass on sandhills.
Trilineata. Common; among grass on sandhills.
Variegata. Frequent; among grass on sandhills.

GENUS WALCKENAERA.

Aggeris. Common; at bottom of rubbish and grass on dry bank sides, near Churchtown.
Monoceros. Very rare; among grass and moss on sandhills.
Fastigata. Very rare; among grass and moss on sandhills.

GENUS PACHYGNATHA.

Clerckii. Frequent; under the ha-ha wall, Formby Parsonage.
Degeerii. Frequent; among grass in sandhills, and on roads, etc., in spring.

FAMILY EPEIRIDÆ.

GENUS EPERA.

Quadrata. Frequent; on bushes, etc.
Apoclisa. Common; on herbage, etc., at edges of dykes.
Solers. Very rare; among dwarf willows on sandhills.
Similis. Common; in balconies, windows, and greenhouses.
Calophylla. Occasional; on bushes and dwarf willows, etc.
Cucurbitina. Rare; on bushes and dwarf willows, etc.
Inclinata. Very common; everywhere.
Diadema. Very common; everywhere.

GENUS TETRAGNATHA.

Extensa. Frequent; among herbage in damp places and over water, etc., stretched at full length in its web.

The name Crustacea is derived from *Crusta*, a crust or hard shell. The animals, which are annulose or articulated, with jointed legs, possess a double or complete circulatory system, and respire by means of bronchiæ, or gills. The external shell, like that of insects, is composed of a dense horny substance called chitine, often strengthened, as in the crab and lobster, by the deposition of carbonate of lime.

The body, being jointed, possesses considerable freedom of motion. The typical number of rings is twenty-one, but these are often soldered together, as we may observe in the crab, so that their relations are obscured. The animal has the power of casting the shell at intervals, and renewing it as the increased growth of the body requires. Otherwise, from the unyielding nature of the carapace, it could not grow.

SPECIES OCCURRING AT SOUTHPORT.

Long-legged Spider Crab (*Stenorhynchus Phalangium*). This curious species is sometimes found on the sands near low water mark; all the specimens we have found have been weakly and damaged, though living.

Slender Spider Crab (*Stenorhynchus tenuirostris*). Bears considerable resemblance to the preceding, but is more slender in its parts, and has little pubescence on the legs. Less abundant than the former.

Scorpion Spider Crab (*Inachus Dorsettensis*). In this species the rostrum is much shorter than in the two preceding, and the hue is much duller.

Hyas araneus. A specimen was once found near where the Whitworth guns were placed. When the legs were extended it covered a space of four inches and a half by three in width. The colours were not obscured, though it had various shells and zoophytes adhering to it.

Harbour Crab (*Carcinus Mœnas*). This is the most abundant kind found on the shore; sometimes eaten by the poor; though small, the flavour is good.

Portumnus variegatus. The shell of this species may be found on the sands, but we have not seen it in the living state.

Velvet Swimming Crab (*Portumnus puber*). Rare; only one specimen, and that in an exhausted state, has been found to my knowledge.

Cleansing Swimming Crab (*Portumnus Depurator*). Common; very active, and swims with great rapidity, burying itself in the sand as the water recedes.

Common Pea Crab (*Pinnotheres Pisum*). Resides in shells, oysters, scallops, cockles, &c. The sexes vary much in appearance, and have, until lately, been considered as distinct species; the female is the *P. varians* of authors.

Angular Crab (*Gonoplax angulata*). This rare species was found by Mr. Graves beyond the end of the Pier. The colours were brighter than in most other kinds.

Masked Crab (*Corystes Cassivelaunus*). A common species, and may be found at most seasons. In the female the front legs are less than half the length of those of the male.

Common Hermit Crab (*Pagurus Bernhardus*). The most abundant species on our shore, generally inhabiting the shell of the common whelk; when left dry it contrives to turn the mouth of the shell downwards. It is very pugnacious; we are unacquainted with its enemies, but have met with great numbers with the abdomen and all the posterior parts eaten away.

Pagurus ulidianus? We name this species with some doubt, having met with numerous specimens inhabiting the shells of *Natica monilifera*, which have a strong resemblance to the figure of this species in Bell's "British Crustacea."

Norway Lobster (*Nephrops Norvegicus*). This beautiful species is given on the authority of Mr. James Glover.

Common Shrimp (*Crangon vulgaris*).

Common Prawn (*Pandalus annulicornis*). Occasionally taken by the shrimpers, but not common, and much smaller than on the south coasts.

Minute Porcelain Crab *Porcellana longicornis*). This minute species is often found on the sponge-like base of the Lobster's-horn Coralline; it is obtained from the size of mustard seed to a quarter of an inch in diameter, and varies in colour from a dull pale red to a brilliant scarlet, intermixed with golden yellow.

Mysis Chamæleon. A specimen was found in 1861.

Pychnogonium littorale. A suctorial crustacean. J. G.

CHAPTER VIII.

I care not, Fortune, what you me deny;
You cannot rob me of free nature's grace;
You cannot shut the windows of the sky
Through which Aurora shows her brightening face;
You cannot bar my constant feet to trace
The lonely shore at dewy morn and eve.
Let health my nerves and finer fibres brace,
And I their toys to the great children leave;
Of nature, feeling, virtue, nought can me bereave.

THOMSON.

MOLLUSCA OF SOUTHPORT.

THE Mollusca are destitute of internal skeleton, and have soft bodies, often protected by an external shell, as in the banded snail of our sandhills (*Helix nemoralis*), and the common cockle. The shell cannot be regarded as essential, for of two species closely allied in structure, *e.g.*, the snail and slug, it is often present in one, and absent or very imperfectly developed in the other.

The Mollusca are further distinguished from the other great sections of the invertebrate division of the Animal Kingdom, the Articulata and Radiata, by the want of symmetry in the two halves of the body, and the absence of joints or articulations, and lateral locomotive appendages.

The majority of our Shells are divided into two classes— Bivalves and Univalves. The Bivalve is a shell in two parts, a right and left valve, connected by a hinge. The Univalve is a conical or spiral shell, often closed by an operculum, which is a plate attached to the foot of the animal, corresponding in shape to the mouth of the shell. "The Mollusca, though nearly all sedentary in their habits, are in their earlier stages swimming animals, being provided with cilia which enable them to move freely about. Aided by these and the ocean currents, they are dispersed, sometimes to immense distances, until they meet with conditions suitable to their growth. It is a remarkable fact that the Bivalves, at this period of their lives, have eyes, to aid them in their movements." Thus there is a natural means by which their over accumulation in any particular part is prevented. After a few days of this free and sportive life, they begin to settle down to the conditions and localities each is destined to occupy. The limpet attaches itself to the rock, between high and low water mark; the cockle, the mya, and the razor-fish bury themselves in the sand and mud; the Teredines attack and burrow into the sides of ships or the hardest wood, and by their silent and ceaseless operations undermine some of the most important works of man; the Pholas excavates itself a home in the rocks and cliffs, by what means science has failed to discover; the mussel forms itself a byssus or cable, by which it is attached to rocks and timber, and one species spins itself a silken nest. Some tribes retain the power of moving about: the Pecten and the Pinna take flying leaps through the water by rapidly opening and closing their valves, the large river mussel pushes itself along with its foot, and the cockle jumps along the sand. The Univalves are provided with a large muscular foot, by which they crawl along the bottom of the sea, or upon aquatic plants and sea-weeds. They have a head, eyes, a mouth armed with jaws, and a tongue, called a lingual ribbon, which is covered with a variable number of minute siliceous teeth. They feed upon confervæ, sea-weeds, and zoophytes; many of them are carnivorous, attacking each other, and also the quiet bivalves. With their file-like tongue they rasp a small hole through the shell, and then devour the helpless inmate; this will explain to the shell-gatherer why so many of the shells he picks up on the shore have little round holes drilled through them. The Bivalves live upon the animalcula and microscopic vegetable matter in the surrounding water; it is carried into the digestive cavities of the animal by currents caused by the action of their ciliary apparatus.

A stranger coming to Southport from one of the inland counties for the purpose of collecting shells, would probably be very much disappointed on his first visit to the shore, as there are certainly very few shells to be found in the immediate vicinity of the town. The best collecting ground is from Birkdale to Formby, at high-water mark, and on the banks near low-water—at high-water mark during the period of the highest tides, especially after heavy west or south-west gales, and near low-water on the slopes of banks during low tides. It is also well to examine the shore at extreme low-water during the times of the highest tides of the year, as at those parts which are not often left uncovered by the water, *Mya truncata* is occasionally found, with its long and curious syphonal tube, alive and perfect. Several minute species may be found by collecting the broken shell and sand from the ripple marks and the slopes of banks, and examining a small quantity at a time in a shallow dish of water, at home. Dredging, unless at a very considerable distance out, is unproductive; scarcely anything can be obtained but species which may be commonly found on the shore.

A collector may make a very pleasant excursion by taking the train to Formby, crossing the sandhills to the shore, about a mile distant, and walking back to Southport, in all from eight to nine miles. A great many shells and other marine curiosities may be found during the walk, and it avoids an otherwise fatiguing return journey.

The Mollusca which have been found on this coast up to the present time, number 146 species. Of these 104 are marine, 7 of them being naked Mollusca (3 Nudibranchs and 4 Sepiadæ); and 42 are land and fresh water shells. The latter not being migratory to any extent, or not so subject to causes of removal, are, of course, actually native, or indigenous to the district. Of the marine species, judging from a lengthened period of observation, 43 may be considered common, or native to the immediate coast; 35 are occasional visitors, living, say within a radius of fifty or sixty miles; and 26 species occur so rarely, live at probably such a distance, and require such conditions, as to oblige us to consider them quite foreign to our shore. These 146 species have been found in a space of seven or eight miles along the shore, and extending about two miles inland. In so small a space, and considering how barren non-observers might suppose the district to be, the collection may be considered a very fair proportion of the whole British Conchology, which comprises about 600 species.

MARINE SHELLS—(BIVALVES).
ACEPHALA LAMELLIBRANCHIATA.

PHOLADIDÆ.

Pholas crispata. Rare; small living specimens have been found in pieces of rotten wood washed up by the tides, and large single valves occasionally. The nearest habitat for this species is Hilbre Island, at the mouth of the Dee, where fine living specimens may be found burrowing in the red sandstone rock at extreme low water.

Pholas candida. Occasionally washed up alive, during very heavy gales; single valves common at all times.

GASTROCHÆNIDÆ.

Saxicava rugosa. Very rare; has been found burrowing in pieces of the zoophyte *Alcyonium digitatum*, and also attached to *Modiola Modiolus* when brought up from deep water by the fishermen.

MYADÆ.

Mya truncata. Not uncommon. Fine specimens may sometimes be taken at extreme low water, during high spring tides.

Mya arenaria. Single valves are not uncommon; perfect shells very rare.

CORBULIDÆ.

Corbula nucleus (the little basket). Occasionally found at high water mark, and on slopes of banks nearer low water. One valve is larger than the other; the smaller one has the appearance of being pressed into the larger, a feature by which the species may be instantly recognised.

ANATINIDÆ.

Thracia phaseolina. One of our most beautiful shells; rather abundant after high tides, and very fine.

Thracia convexa. Very rare, and single valves only.

SOLENIDÆ.

Solen marginatus (Sword Shell). Rare, and generally single valves; the shell is from four to six inches long, three-quarters of an inch broad, with a groove indented at the hinder margin.

Solen Siliqua (Razor Shell). Good perfect shells have been found, but rarely; single valves occasionally. This species sometimes attains a great size, nine or ten inches long, and an inch and a half broad. In many parts of the kingdom it is used as an article of food, and considered very delicate eating; it lives buried in the sand at low water, from one to two feet deep. The creatures are caught by pushing crooked wires down the hole and hauling them up, or a little salt is dropped down the hole, which rather incommodes the animal. It rises up to see what is the matter, and is seized; but if thrown upon the sand will very quickly work its way down again with its powerful muscular foot.

Solen Ensis (the Scymitar). Very abundant and fine. Three to four inches long, and curved like a bow.

Solen Ensis (Var. *magna*). Double the size of the preceding, and very rare.

Solen pellucidus. Found attached to bunches of coralline; rather scarce; from an inch and a half to two inches long, and a quarter broad; the hinge margin straight, the outer margin bowed. A novice would perhaps mistake the young *Ceratisolen Legumen* for this species; the difference is easily known by the position of the hinge; in all the *Solens* it is near one end; in *C. legumen* it is in the centre of the hinge margin.

Solen coarcticus. *Solen candidus.* Single valves of these two latter species, fine and in good condition, have been found.

SOLECURTIDÆ.

Ceratisolen Legumen (the Peas Pod). This one has a long flat shell, as the name denotes; it lives buried in the sand at extreme low water; rather common.

TELLINIDÆ.

Psammobia ferroensis (Sunset Shell). A flat elongated oval shell, rounded at one end, and squarish at the other end, and prettily rayed with pink from the hinge to the front margin. Common as a British species, but rare at Southport. Sometimes found perfect, attached to bunches of corallines.

Tellina tenuis. Common. A very pretty species, the shell variously coloured, rose, pink, yellow, white, etc.; flat or compressed, rounded in front, attenuated behind, about one inch long and five-eighths wide.

Tellina fabula. Abundant; one valve of this species is smooth, the other, upon close examination, will be found to be marked with very fine concentric lines, which cause it to be slightly iridescent. They are mostly about three quarters of an inch long, and half an inch wide, very much compressed, rounded in front, attenuated behind, much more so than *Tellina tenuis*, and nearly white.

Tellina donacina. An elegant oval shell, radiant with the colours of the setting sun ; single valves only have been found.

Tellina solidula, one of our commonest shells, varying in colour through all the shades from crimson to yellow. Roundish and solid, from one quarter to three quarters of an inch in diameter.

Syndosmia alba. A pretty oval, shining, rather pellucid, white shell, moderately plentiful.

Scrobicularia piperata. Not uncommon ; found in the greatest numbers opposite the Promenade, especially after a heavy sea, when it is washed up from the mud in which it burrows. Very flat, round, white, and varies in size from half an inch to an inch and a half in diameter.

DONACIDÆ.

Donax anatinus (Wedge Shell). Common about low-water mark, and often very fine. Although one of our commonest shells it is not the least handsome, being a light olive colour outside, and frequently a brilliant clouded violet within.

MACTRIDÆ.

Mactra subtruncata. Rather common, and generally small. White, solid, and somewhat triangular in shape.

Mactra elliptica. I have found one good perfect specimen only of this shell.

Mactra stultorum. Very common and fine. Sometimes at low-water, lying in groups of hundreds together, within spaces of a few yards. The gulls break immense numbers to get at the animals.

Mactra solida. A few single valves have been found.

Lutraria elliptica. Single valves are occasionally met with; perfect shells very rarely. It is a large, oblong, rather flat shell, gaping or open at the extremities.

Tapes pullastra. Artemis lincta. A few single valves of the two latter species have been found. Species that are so rarely picked up, and then only in single valves, are, as a rule, found attached to corallines.

VENERIDÆ.

Venus striatula. Common. Triangular. Looking at the shell edgeways, it is heart-shaped, has highly raised concentric lines, and is sometimes handsomely marked with rich brown rays.

Venus ovata. Very rare. Similar in shape to a cockle, but small and white, and the sculpture much finer.

Lucinopsis undata. Moderately common. A roundish shell, white, slightly tinged with rust colour; varies from a quarter to three quarters of an inch in length and breadth.

CYPRINIDÆ.

Cyprina Islandica. Perfect shells of this fine species are rarely to be met with; single valves not uncommon.

CARDIADÆ.

Cardium echinatum. A large species of the Cockle tribe, with thick radiating ribs, bristling with tuberculous spines.

Single valves are common; perfect double specimens may sometimes be met with after high tides.

Cardium edule (Common Cockle). A very common shellfish about Southport at all times. In 1858, a cockle-bed or "scour," as it is locally termed, was discovered about five miles north-east of the town, where the cockles were so numerous as to be literally shovelled up with spades; the yield for several months was from ten to fifteen tons a-week. It is scarcely possible to realize the prodigious numbers taken from the bank, as a ton contains about 80,000 individual cockles.

Cardium Norvegicum. A small delicate-looking shell; single valves found upon corallines.

LUCINIDÆ.

Lucina leucoma. Extremely rare; one or two single valves only found.

KELLIADÆ.

Montacuta ferruginosa. Very rare, and single valves. A small white, oval, semi-transparent shell, about three-sixteenths of an inch long, generally stained with rust-colour. This and the following species are found by collecting and washing the sand and broken shell from the ripple-marks and slopes of banks.

Montacuta bidentata. Moderately common A minute white, oval, almost transparent shell; about one-eighth of an inch long.

MYTILIDÆ.

Mytilus edulis (Common Mussel). Common. Sometimes in great numbers attached to pieces of wood or sea-weed.

Modiola modiolus (Horse Mussel). Sometimes brought up by the fishermen, and frequently very large, from five to six inches in length.

Modiola tulipa. This is a pretty translucent radiated shell. Good living examples have been found at low water, opposite Birkdale.

Crenella discors. Rare. Has been found at low water, burrowing in *A. digitatum*.

ARCADÆ.

Nucula nucleus. Not common. Generally single valves attached to bunches of coralline. A small dull olive-coloured shell, the inside pearly white, and about twenty minute teeth on the margin at one side of the hinge, and ten at the other side.

Leda caudata. Several specimens have been found on the Birkdale shore.

OSTREADÆ.

Pecten maximus (the Great Scallop). Great numbers of this, our largest British bivalve, are brought up by the fishermen.

Pecten opercularis (the Common Scallop, or Fan Shell). Not uncommon, but generally small; a handsome species, varying much in colour, being sometimes yellow, orange, crimson, brown, purple, white, or mottled.

Pecten varius. Not commonly found, but on one occasion ten or a dozen were picked up.

Ostrea edulis (the Oyster). Not common on the shore; occasionally brought up by the fishermen.

Anomia ephippium. Occasionally found upon *Modiola modiolus.* This curious mollusk is attached to shells, rocks, and stones by a muscle projected through an orifice in the lower valve, near the hinge; it is a lustrous, pearly shell, and adapts itself to the shape of the body to which it is attached.

Anomia patellaformis. Anomia aculeata. Like the *ephippium,* these two latter have been found on *Modiola modiolus.*

UNIVALVES.

GASTEROPODA NUDIBRANCHIATA.

EOLIDIDÆ.

Œolis coronata.
Œolis papillosa.

DORIDIDÆ.

Doris Johnstoni.

The *Nudibranchiata* are the naked mollusks or sea-slugs. They are curious, and some of them beautiful animals. We owe their presence to the construction of the Pier, as they are found on its timbers and on the stones heaped round their foundations.

GASTEROPODA PROSOBRANCHIATA.

PATELLIDÆ.

Patella vulgata (Common Limpet). Rare, and when found very much worn.

Patella athletica. Like the above, very rare.

DENTALIADÆ.

Dentalium entalis (Tooth Shell). Some years ago this was a common shell, but now is only occasionally found, and generally attached to corallines. Fine specimens are about an inch and a half long, tubular, tapering to the posterior end, and slightly curved; perfectly white.

Dentalium Tarentinum. The same remarks apply to this species as to *D. entalis*, with the exception of the specific differences. It is a thicker and straighter shell, the posterior end marked with very fine raised lines, lengthwise, so fine as to require the aid of a microscope to discover them; they are the principal characteristic mark of the species.

Pileopsis Hungaricus. Shaped like the conventional cap of Liberty. Many years ago, when tides were deep on the foreshore, this was sometimes found, but is now extremely rare. Alterations in channels and sandbanks influence the occurrence of shells.

FISSURELLIDÆ.

Emarginula reticulata. Rare. Found at high-water mark, and amongst corallines. In shape like a cap of Liberty, with a slit in the front margin.

TROCHIADÆ.

Trochus zizyphinus (Top Shell). Moderately common. A cone-shaped shell, granulated in narrow spiral bands.

LITTORINIDÆ.

Littorina littorea (Periwinkle). One of the most abundant shells on rocky coasts. Formerly uncommon at Southport, but since the construction of the Pier may be found in abundance.

Littorina rudis. Rare. A smaller and lighter coloured shell than the last.

Littorina littoralis. Good fresh specimens have been found.

Rissoa vitrea. Rare. Minute, shining white. Found by collecting and washing sand, as previously stated.

Rissoa ulvæ. Very common. In walking along the shore we frequently see patches of what the stranger would suppose to be black sand; if a portion of it be taken up it will be found to be a mass of these small shells.

Rissoa castanea. Rather rare. Found amongst *R. ulvæ*, and similar in shape and colour, but very much larger. I named this shell what I believed it to be; Jeffries has examined it, and says it is an unrecognised variety of *ulvæ*.

TURRITELLIDÆ.

Turritella communis (the Common Cockspur). Very common on most parts of the shore.

Cæcum glabrum. A minute shell found in the fine shell *débris* and sand which I have taken home to dry in order to search for Foraminifera.

CERITHIADÆ.

Apporhais pes-pelecani (the Bird's-foot Shell). A handsome and not uncommon species. The lip is extended out in such a way as to resemble a bird's webbed foot, from which its specific name is derived.

Cerithium reticulatum. A rather worn specimen found when searching for Foraminifera

SCALARIADÆ.

Scalaria Turtonis (Wentle-trap, or Double Cockspur). Not uncommon. A handsome shell, turreted in shape, the whorls round and distinct, and crossed lengthwise with rather flat, pale brown, moderately close ribs.

Scalaria communis (Common Wentle-trap). Moderately common. Same shape as the preceding, but the whorls more distinct, and the ribs thicker and more prominent.

PYRAMIDELLIDÆ.

Aclis supranitida. Rare. A pretty but very small shell, from one-tenth to three-tenths of an inch long, conical or turreted in shape. Good specimens are ornamented with raised spiral lines or ridges. It is considered rather rare as a British species. Found by collecting and washing the sand as previously stated.

Aclis ascaris. Also one of the rare minute species, and found with the above.

Eulima polita. Very rare. Lanceolate or tapering in shape, being about five-eighths of an inch long and one-eighth broad at the base; colour, a shining porcelain white.

Eulima subulata. Not uncommon. Very narrow, finely tapering, light brown, with spiral bands of a darker shade.

Chemnitzia elegantissima. Very rare. A minute, white, spiral shell, with elevated oblique ribs on the whorls. Sometimes found along with *A. supranitida*.

Chemnitzia rufa. Found with the above.

Odostomia interstincta. Rare. Found with the above. A minute species, requiring the aid of the microscope for identification. It is perfectly white, with longitudinal ribs upon the whorls.

Odostomia indistincta. Odostomia rissoides var. dubia. These, like previous minute species, are rare, not only on the Southport shore, but generally; they are found when searching for Foraminifera, most of them being difficult to identify with certainty. Were named for me by our great authority, Mr. Jeffries.

NATICIDÆ.

Natica monilifera. Common. A handsome shell, being globular in shape, highly polished, and ornamented with a spiral band of brown spots.

Natica nitida. Not common. Similar to *N. monilifera*, but smaller, about a quarter of an inch in diameter, and not quite so globular, the spiral a little more produced.

MURICIDÆ.

Murex erinaceus (Sting Winkle). Not uncommon, though generally rather small.

Nassa incrassata (Dog Whelk). Rare, though common as a British species. In shape it is similar to the common whelk, about half an inch long, with thick longitudinal ribs.

Purpura lapillus. Not common. Being naturally an inhabitant of rocky localities, the specimens are often much worn when they arrive upon the Southport shore. It is fusiform in shape, very solid, and about an inch long. The animal secretes a milky fluid, which in former times was used in the production of a rich purple dye.

Buccinum undatum (Common Whelk). Moderately plentiful; very fine specimens may sometimes be found after storms. This is a common shell all round the British coasts. In many parts it is taken in great numbers, and used for bait; and quantities are sent to the London markets, where they are boiled and eaten.

Fusus Islandicus. Very rare: generally weather-worn.

Fusus antiquus. Common after heavy gales, in company with the whelks, from which it may be known by the canal being more elongated, and the shell generally smoother and more tapering. The fishermen sometimes bring up splendid specimens from deep water, measuring six to seven inches long, and perfectly white.

Trophon muricatus. Very uncommon. I have only found one good specimen.

CONIDÆ.

Mangelia gracilis. Very rare; one or two specimens have been found near low water.

CYPRÆADÆ.

Cypræa Europœa (Cowrie). Not common. It is about the size and shape of a coffee berry, with raised lines or ribs across; a pale flesh colour.

GASTEROPODA OPISTHOBRANCHIATA.

The Mollusks of this order may be termed sea-slugs, since the shell, when it exists, is usually small and thin, and wholly or partially concealed by the animal.

BULLIDÆ.

Cylichna cylindracea (the Paper Roll). So named from its shape. Rare on the Southport coast. About half an inch long, and three-sixteenths of an inch wide; white and shining.

Cylichna obtusa. Similar to the above, but half the length. A few years ago this shell was plentiful close to the town;

it is now more abundant four or five miles to the west, at high-water mark.

Tornatella fasciata. Very abundant. Not unlike a shuttle in shape, but broader in proportion, and beautifully coloured with bands of pink and white.

Scaphander lignarius. Many years ago this was not an uncommon shell on our shore ; it is now extremely rare.

Philine aperta. Common. An extremely thin, white, translucent shell, without spire, and a wide open mouth.

CEPHALOPODA DIBRANCHIATA.
SEPIADÆ.

Sepia officinalis (the Common Cuttle-fish). The internal shell of this mollusk is occasionally washed up in considerable numbers during heavy gales in the winter. It is six to eight inches long, three inches in width, oval, and extremely light in proportion to the bulk. The class *Cephalopoda* ranks the highest in the mollusca, as in the complexity of its organisation it approaches most nearly to the vertebrated animals. It is named from the locomotive organs being arranged round the head ; when in the water, or crawling amongst rocks or on the strand, the animal has the appearance of being head downwards. The Sepiadæ have eight short lanceolate, and two long tentacular arms. The large and prominent eyes are situated underneath the arms, one on each side ; above, in the centre of the circle of arms, is a strong horny beak. The arms and tentacula, besides being organs of locomotion, serve to catch and hold their prey ; and, as they are covered with small suckers, they are enabled to maintain so tenacious a hold that any unfortunate crab or fish with which they come in contact is left without escape. Whilst possessing such powers of offence, they are gifted with most singular means of defence. Like the chameleon, they have the power of changing their colour to delude their foes; they also possess a bag, from which, when pursued, they eject a quantity of inky fluid, which envelopes them in a black cloud, and covers their escape. The contents of the ink-bag supply the brown pigment called sepia, used by artists.

Sepia biserialis. A single specimen of the internal shell or bone of this rare species has been found on the Southport shore. It is much smaller than *S. officinalis*, lanceolate in shape, the point curved a little outwards, and the base slightly inwards.

Sepiola Atlantica. A much smaller animal than the common Cuttle. I have seen only one good specimen found on the shore.

Voligo vulgaris. A fine specimen has been taken at the end of the Pier.

LAND AND FRESH-WATER SHELLS.
ACEPHALA LAMELLIBRANCHIATA.
CYCLADIDÆ.

Cyclas rivicola (River Cycle or Fresh-water Cockle). Found in many of the streams and ditches about Southport, but small; in the canal at Burscough Bridge, abundant and fine. In shape this shell is similar to a young cockle, but more compressed; finely striated, greenish brown in colour, with a narrow yellow band round the margin.

Cyclas cornea (Horny Cycle). In almost any ditch. A round and dumpy shell, generally dark brown, varying in size from one-eighth to half an inch in diameter.

Cyclas calyculata (Capped Cycle). Found sparingly in a stream by the first bridge beyond Churchtown, along the road to Martin Mere. A very transparent shell, about a quarter of an inch long; the umbones—the parts above the hinge—very prominent.

Pisidium amnicum (River Pera). Not uncommon in ditches on the Moss, and on the Martin Mere road, beyond Churchtown, but very small, rarely measuring more than one-eighth of an inch long; common and very fine in the canal at Burscough Bridge. In shape obliquely oval, with minute raised ribs.

UNIONIDÆ.

Anodonta cygnea (Swan Fresh-water Mussel). Common in many of the streams and ditches on the Moss, varying from two to four inches in length, oval in shape, olive or brown outside, bright pearly within. One of the largest and handsomest of the British shells.

GASTEROPODA PROSOBRANCHIATA.

PALUDINIDÆ.

Paludina Listeri (Marsh Shell, or River Snail). Not uncommon in many of the ditches on the Moss. At the approach of winter it buries itself deep in the mud, and makes its appearance again with the warm days of April. A very handsome shell, transparent horn colour, with three dark brown spiral bands. Fine specimens are an inch and a half high, and an inch and a quarter wide at the base, with five or six very convex volutions. When the animal is retracted it is closed by an operculum, a sort of trap door, which should always be fitted in the shell when the animal is taken out. It is both useful and ornamental in an aquarium, the animal being beautifully sprinkled with golden spots, and feeding mostly on the confervæ growing upon the sides of the glass.

Bithinia tentaculata (Tentacled Bithinia). Common in most of the ditches upon the Moss; about a quarter of an inch wide at the base and half an inch high, with five rather flat volutions of dark brown or yellowish horn colour, the aperture closed by an operculum.

Valvata piscinalis (Stream Valve-shell). Shell globular, with an elevated obtuse spire, the volutions well rounded and distinct; the aperture closed by a valve or lid. Common in the large drains upon the Moss.

Valvata cristata (Crested Valve-shell). Not uncommon in the same situations as the preceding, but very different in appearance, being not more than an eighth of an inch in diameter; discoid, flat above, and concave or umbilicate beneath.

ASTEROPODA PULMONIFERA.
LIMACIDÆ.

Arion empiricorum (Black Arion). Common in the fields after rain and in damp weather; as moisture is an absolute necessary of the creature's existence, it is rarely seen in very dry weather. This is the common jet black snail;

it varies in colour according to locality. It is found in woods, of many colours—white, yellow, orange, and reddish brown. It is essentially a vegetable feeder, but will sometimes regale itself with a dead worm. It has no distinct shell.

Limax agrestis (Milky Slug). A small dark, or reddish grey, and voracious vegetable feeder; common in fields, upon hedge-banks, and in gardens. The shell is a small squarish oval, white, calcareous plate, slightly convex above, situated underneath the skin of the shield, a little behind the head. The animal, when extended, measures from an inch to an inch and a half in length, and when irritated pours out a white milky fluid from the pores all over its body.

Limax cinereus. Not so common as the preceding; found amongst grass in damp situations, under logs of wood, about outhouses and gardens. It is a large, dark grey slug, sometimes nearly black, measuring from three to five or six inches long, and proportionately bulky; the back and tail coarsely wrinkled, and the mucus colourless. The shell is internal, from a quarter to half an inch long, half that width, slightly convex, rather pearly white, and sometimes tinged with pink.

HELICIDÆ.

Vitrina pellucida (Transparent Glass-bubble). Small, extremely thin and transparent, very highly polished, and of a pale watery green colour. Not uncommon amongst moss and under stones upon the Birkdale sandhills.

Zonites alliarius (Garlic Snail). Found under stones upon the sandhills, but rather rare. About a quarter of an inch in diameter; the upper side slightly convex, very bright, shining, rather transparent, yellowish horn colour. This species is easily recognised, as when the animal is irritated by touching it emits a strong odour of garlic.

Zonites nitidulus (Dull Snail). Small and very rare. Has been found amongst moss on the hills at the end of "Peter's slack."

Zonites nitidus. One of our rarest shells; has been found in Birkdale.

Zonites purus (Delicate Snail). Very rare; amongst moss on the Birkdale sandhills. Small, about one-sixth of an inch in diameter, depressed, transparent, yellowish white, rather shining, and slightly wrinkled.

Helix nemoralis (Girdled Snail). One of our commonest shells, being found everywhere upon the sandhills; at the same time it is the most beautiful. Very variable in the colour and markings, being sometimes white, yellow, pink, reddish, or brown, or marked with five or fewer rich chocolate bands.

Helix caperata (Black-tipped Snail). Common in Birkdale, particularly on the sandhills between the two roads at the entrance to the Park.

Helix hispida (Bristly Snail). Sparingly found in the hedges of the fields near the Rectory, also in Birkdale. It is about a quarter of an inch in diameter, five or six whorls, slightly convex, horn coloured, and covered with very fine short bristles.

Helix pulchella (White Snail). Not uncommon upon the sandhills; generally and usually among moss. A beautiful little shell, less than one-eighth of an inch in diameter;

a pure opaque white, and sometimes brownish, rather flat above, a small umbilicus beneath, the mouth reflexed, and a little thickened round the margin.

Helix pulchella (var. *costata*). Found with the preceding. The same size and shape, of a pale brown tint, and ornamented with raised radiating ribs visible only under a magnifying lens.

Helix rotundata (Radiated Snail). Very rare; among grass and under stones upon the sandhills. Nearly a quarter of an inch in diameter, the under side almost flat; grey, with dark brown spots.

Helix pygmæa (Pigmy Snail). Very rare; found in damp situations, under pieces of wood or stones. Very minute, convex on both sides; shining, brown, semi-transparent.

Pupa muscorum (Margined Chrysalis-shell). Moderately common; among moss and low plants upon the sandhills. As the name denotes, the pupæ are shaped like a chrysalis. This species is about one-eighth of an inch high; shining, dark brown, the margin a little reflected, with a thick white band round the outside, and a single minute tooth in the centre of the aperture on the body whorl.

Pupa edentula (Toothless Whorl-shell). Rather scarce. Found in the same situations as the preceding; about the same size and colour, but without tooth; the edge of the aperture or lip simple, without margin or rib.

Clausilia laminata. A dead specimen found in Birkdale; it must have been introduced, probably with plants or shrubs received from a distance.

Zua lubrica (Common Varnished Shell). Not uncommon in the sandhills and fields behind the Rectory, but oftener dead and eroded than living. The shell is about a quarter of an inch high, cylindrical oblong, reddish brown, very bright and glossy.

Succinea putris (Common Amber-shell). Rather abundant in a small watercourse on the Birkdale sandhills, half a mile beyond the church, and under the bridge by the boundary stone, on the Scarisbrick-road.

LIMNÆADÆ.

Physa fontinalis (Stream Bubble-shell). Rather plentiful in the "river Nile," and in the streams on the Moss. The mouth of this shell opens to the left hand, by which characteristic it is easily recognised; it is very thin, transparent, brown, and highly polished.

Physa hypnorum (Slender Bubble-shell). In the same localities as the above, but not quite so frequently. It has the same characteristics, with the exception of being longer and narrower in proportion.

Planorbis albus (White Coil-shell). Moderately common in the stream by the first bridge on the Martin Mere road; dark coloured, finely striated, concave underneath, slightly so above; the outside coil rapidly enlarging.

Planorbis vortex (Whorl Coil-shell). Common in many ditches and stagnant pools; a thin light-brown shell, with six or seven volutions, about three-eighths of an inch in diameter, flat above, and slightly concave beneath.

Planorbis spirorbis (Rolled Coil-shell). Found with *P. vortex*, but not so commonly. Light brown, slightly concave on both sides, with six volutions.

Planorbis nautileus. A good specimen found in a ditch in Birkdale.

Planorbis contortus (Twisted Coil-shell). In many ditches on the Moss. The whorls are very closely coiled and very narrow, the upper surface rather flat, the underneath deeply concave.

Limnæus pereger (Puddle Mud-shell). In every ditch, pond, and stream, where it may be commonly found crawling about the bottom.

Limnæus stagnalis (Lake Mud-shell). Fine specimens may be found in ditches in Birkdale Park, and in ditches on the Scarisbrick-road and on the Moss. It is a handsome shell, light brown, an inch and a half high, the body whorl large and open, the spire of six or seven volutions tapering to a fine point.

Limnæus glaber. Also found in Birkdale ditches, but very rare.

Limnæus truncatulus (Ditch Mud-shell). Not uncommon in most ditches on the Moss, generally at the surface of the water, close to the side. Shell dark brown, about a quarter of an inch high.

Limnæus palustris (Marsh Mud-shell). Rather common in ditches on the Moss, in Birkdale, and in the neighbourhood of Churchtown. Shell dark brown, about three-quarters of an inch high, body whorl longer than broad, the spire gradually tapering to a point.

Ancylus fluviatulus (Common River Limpet). Very rare. Found on stones in a stream on the Moss. The shell is about a quarter of an inch in diameter and height, cone-shaped, with the apex curved backwards and near one end. Semi-transparent, light greenish horn colour; inside blueish white, shining.

CHAPTER IX.

> And here were coral bowers,
> And grots of madrepores,
> And banks of sponge, as soft and fair to eye
> As e'er was mossy bed
> Whereon the wood-nymphs lie
> With languid limbs in summer's sultry hours.
>
> SOUTHEY.

ZOOPHYTES OF SOUTHPORT.

THE term Zoophyte is applied to all those productions which, bearing a strong resemblance to vegetables in form and some other particulars, are yet of an animal nature. The arborescent forms are often called Corallines, a name particularly appropriate, being a derivative of the word Coral. They are intimately allied to the Corals by means of which such gigantic changes are daily being effected. Islands and continents are being raised from the deep abysses of the ocean, to be hereafter clothed with vegetation and probably made the seat of a busy population—and these mighty results are being brought about by the agency of minute creatures, scarcely perceptible to our unaided sight, but whose operations, though slow, silent, and invisible, are yet certain and increasing :—

> Unconscious, not unworthy, instruments,
> By which a hand invisible was rearing
> A new creation in the secret deep.
> Omnipotence wrought in them, with them, by them;
> Hence, what Omnipotence alone could do,
> Worms did. I saw the living pile ascend,
> The mausoleum of its architects,
> Still dying upwards as their labours closed;
> Slime the material, but the slime was turned
> To adamant by their petrific touch;
> Frail were their frames, ephemeral their lives—
> Their masonry imperishable.
> MONTGOMERY.

Amongst the many recent cultivators of this interesting department of natural history, the name of the late Dr. Johnston, of Berwick, stands pre-eminent; his excellent work on the British Zoophytes has done much to exalt the subject and to diffuse a more general taste for its cultivation.

"Zoophytes," to adopt the language of Dr. Johnston, "present to the physiologist the simplest independent structures compatible with the existence of animal life, enabling him to examine some of its phenomena in isolation, and free from the obscurity which greater complexity of anatomy entails. The means of their propagation and increase are the first of a series of facts on which a theory of generation must arise; the existence of vibratile cilia on the surface of the membrane, which has since been shown to be so general and influential among animals, was first discovered in their study, and in them is first detected the traces of a circulation carried on independently of a heart and vessels. The close adhesion of life to a low organisation; its marvellous capacity of redintegration; the organic junction of hundreds and thousands of individuals in one body, the possibility of which fiction has scarcely ventured to paint in its vagaries, have all in this class their most remarkable illustration."

Not much more than a century has elapsed since the true nature of these productions was first discovered; prior to that period various opinions were entertained respecting them. By one class of persons—and these were by far the most numerous—they were regarded as undoubted subjects of the vegetable kingdom, and were so arranged and classified in the various systems of the most learned botanists of the day. Nor is this to be wondered at when we consider the striking resemblance which these objects bear to vegetables both in form and habits; some of them being eminently arborescent in their mode of growth, and fixed by a kind of root, either embedded in the sand, or attached to rocks, stones, and other substances, in the same manner as sea-weed, and consequently incapable of locomotion, except during the brief period of their embryonic life, a character formerly considered essential to the idea of an animal, locomotion being common to all the animals then known.

By a second set of observers, at the head of whom stood the illustrious Linnæus, all the horny and flexible zoophytes were considered to hold a station intermediate between the animal and vegetable kingdoms, partaking of the nature of both. The Lithophyta, however, were placed by him in the animal kingdom, on the supposition that lime was always an animal product. "The animalcules of the Lithophyta, like the testaceous tribes," he said, "fabricate their own calcareous polypidom, forming the whole mass into tubes, each ending on the surface in pores or cells, where alone the animal seems to dwell; but the polypes of the proper zoophyta, so far from constructing their plant-like polypidoms, are, on the contrary, the productions or efflorescences of it; just as the flowers do not make the herb or tree, but are the results of the vegetative life proceeding to perfection." Polypes, according to this fancy, bore the same relation to their polypidom that flowers do to the trunks and branches of a tree—both grew by vegetation; but while the one evolved from the extremities blossoms which shrank not under external irritation and were therefore properly flowers, the other put forth flowers which, because they exhibited every sign of animality, were therefore with reason considered animals. In a letter to Ellis he remarks, alluding to the zoophytes, "they are, therefore, vegetables, with flowers like small animals." In his "Diary," Linnæus further remarks that they are "vegetables with respect to their stems, and animals with respect to their florescence."

Zoophytes were deemed by other naturalists to be of mineral origin. This theory was particularly advocated by Henry Baker. "The rocks in the sea on which these corals are produced," he says, "are undoubtedly replete with mineral salts, some whereof, near their surface, being dissolved by the sea water, must consequently saturate with their saline particles the water round them to a small distance, where, blending with the stony matter with which the sea water always abounds, little masses will be constituted here and there and affixed to the rocks. Such adhering masses may be termed roots, which roots, attracting the saline and stony particles, according to certain laws in nature, may produce branched or other figures, and increase gradually by an apposition of particles becoming thicker near the bottom, where the saline matter is more abounding, but tapering or diminishing towards the extremities, where the mineral salts must be fewer in proportion to their distance from the rock whence they originally proceed; and the different proportions of mineral saline particles of the stony or other matter wherewith they are blended, and of marine salt, which must have a considerable share in such formations, may occasion all the variety we see. Nor does it seem more difficult to imagine that the radiated, starry or cellular figures along the sides of these corals, or at the extremities of their branches, may derive their productions from salts incorporated with the stony matter, than that the curious delineations and appearances of minute shrubs and mosses on slates, stones, etc., are owing to the shootings of salts intermixed with mineral particles; and yet these are generally allowed to be the result of mineral steams and exhalations."

It is scarcely necessary to observe that all these theories, however ingenious and interesting, are untenable; the beautiful and poetic hypothesis of Linnæus is, however, the nearest approximation to the truth. We learn from Dr. Johnston's "Introduction to the British Zoophytes," on the authority of M. de Blainville, that Ferrante Imperato, an apothecary at Naples, was the first naturalist distinctly to announce, as the result of his own observations, the animality of corals and madrepores. He is said to have added to the description of the species which fell under his notice, illustrative figures of considerable accuracy, although the "Historia Naturale" was published so early as the year 1599.

This discovery, however, had no result, since there is evidence of its entire rejection and ultimate neglect by those who studied nature. It is to John Ellis, a London merchant in the middle of the last century, that we are indebted for having placed the animality of zoophytes beyond all doubt or controversy. "There was nothing unformed or mystical in Ellis's opinion. Certain marine productions, which, under the names of Lithophyta and Ceratophyta, had been arranged among vegetables, and were still very generally believed to be so, he maintained and proved, with a most satisfactory fulness of evidence, to be entirely of an animal nature, the tenements and products of animals similar in many respects to the naked fresh-water polype. By examining them in a living state through an ordinary microscope, he saw these polypes in the denticles or cells of the Zoophyta; he witnessed them display their tentacula for the capture of their prey; their varied actions and sensibility to external impressions and their mode of propagation; he saw, further, that these little creatures were organically connected with the cells, and could not remove from them, and that although each cell was appropriated to a single individual, yet was this united by a tender, thready line to the fleshy part that occupies the middle of the whole coralline, and in this manner connected with all the individuals of that coralline. The conclusion was irresistible: the presumed plant was the skin or covering of a sort of miniature hydra,—a conclusion which Ellis strengthened by an examination of their covering separately, which he said was as much an animal structure as the nails or horns of beasts, or the shell of the tortoise; for it differs from sea-plants, properly so called, such as the Algæ, Fuci, etc., which afford in distillation little or no traces of a volatile salt; whereas the corallines afford a considerable quantity, and in burning yield a smell somewhat resembling that of burnt horn and other animal substances, which of itself is a proof that this class of bodies, though it has the vegetable form, yet it is not entirely of a vegetable nature."

It would be foreign to the nature of this work to enter into the minute anatomy, development, or classification of the various tribes of zoophytes. The subject has been investigated with great industry and success by recent observers, and for a summary of our present knowledge we would refer students to Dr. Carpenter's work on the Microscope, Landsbrough's "Popular History of British Zoophytes," and the classical work of Dr. Johnston.

Spongia mammillaris (Nipple Sponge). We were much gratified in obtaining this interesting species in a living state, left on the sands after a very heavy gale of wind. It continued to eject water from the summits of the projecting parts for several days after it was found. It is the only sponge we have been able to procure on this shore retaining its vitality.

Hydractinia echinata. A very common and abundant species, found coating a variety of old as well as living shells, on which it is most abundant, such as *Buccinum undatum*, (the Common Whelk), and the *Natica monilifera*. It has been stated that most of the shells encrusted by this polype are tenanted by the Hermit Crab; we have frequently found them so, but it does not hold good as a general rule; in several instances we have found this species growing on *Mya truncata*.

Coryne pusilla? On sea-weeds, old shells, and frequently on other zoophytes. The species are all very small, and are only accidentally to be met with.

Eudendrium rameum. Frequently thrown ashore after heavy gales, adhering to old shells, stones, and occasionally on the stems or roots of the larger kind of sea-weed.

Eudendrium ramosum. This and the last species are rarely obtained with living polypes, except from deep water, in five to twenty fathoms. It is often brought up in the trawl nets.

Tabularia indivisa (Tabular Coralline). This curious species inhabits deep water, and is commonly thrown ashore attached to stones and shells; we have frequently obtained specimens with the living polypes on by following out the receding tide. After strong winds it is to be met with in great abundance on the shore. It is also obtained by dredging in from five to twenty fathoms, with the living polypes, and may be kept alive in the aquarium for a considerable time, if well supplied with sea water. It grows from three or four inches to a foot or more in height. It is of a dull horn colour, and occurs mostly in dense clusters. The polypes are of a bright red colour, they are usually thrown off after a few days' continuance, and are soon renewed. The fabric varies considerably in general appearance, being in some instances quite straight and entire, in others much curved and branched.

Tabularia Laryna. A very delicate species, and much clustered; of a lighter colour than the preceding, but smaller and more transparent. Is found at times in abundance, on the rejectamenta left by the receding tide.

The *Laryna* forms tufts from one to two or three inches in height, and to obtain the polypes alive must be procured from deep water.

Tabularia gracilis. A beautiful species, much resembling *T. indivisa*, but more slender. The polypes are larger and brighter coloured than in the latter. It usually attains the height of three or four inches, growing on other species of zoophytes, and is only to be obtained alive from deep water. It is not abundant here, though, like numerous other kinds, it is cast ashore during severe weather.

Helicina helicinum (Herring-bone Coralline). This beautiful species is to be met with in very considerable abundance at certain seasons, but is only found after rough weather; in May and September we have found it with the living polypes, adhering to shells and stones. It is so much like a diminutive dry tree that it is frequently passed as a decaying vegetable. After stormy weather it may be found in considerable quantities from Formby to Crossens.

Sertularia polyzonias. A very pretty and not uncommon species, affording a beautiful microscopic object. Mostly found on stones, shells, and sea-weed. It varies from one to several inches in height, and is of a pale fawn colour, with the vesicles of a clear colourless substance, giving out prismatic reflections. We have found it lining the inside of the *Cardium aculeatum*.

Sertularia rosacea (Lily or Pomegranate-flowered Coralline). On shells, stones, corallines, and sea-weeds. A most beautiful species, and found in tolerable abundance on the receding of the tide. It is from one to two inches in height, extremely delicate and slender, and of a dusky straw colour, sometimes with a rosy tint. It creeps along the substance on which it grows, and is to be met with on the sands from Formby to Crossens.

Sertularia abietina (Sea Fir). One of our more beautiful zoophytes. It is very abundant; parasitic on stones and shells in deep water, and after high tides or stormy weather is thrown in great quantities on the shore, but rarely obtained with living polypes, except when dredged up from deep water. Frequently quite encrusted with serpulæ and small mussels; at times coated with various species of Lepralia and Celepora.

Sertularia fallax. A small but very elegant species. Not abundant, but to be found attached to oysters and scallops, and left ashore by the retreating tide. A native of deep water.

Sertularia tamariscina (Sea Tamarisk). Frequently found after spring tides or rough stormy weather, attached to shells and stones; at times forming clusters eight or ten inches in height. When recently left by the tide and with polypes living, the general colour is bright amber, but soon changes to a dull brown. An inhabitant of deep water, and occasionally brought up in the dredge-net. We sometimes see it exposed in the market, growing on oysters.

Sertularia filicula (Fern Coralline). This common, but beautiful species, is frequently found with its polypes alive, growing on sea-weed, and, like the last species, is much encrusted with Lepralia and Serpulæ; it will live for a considerable time if well supplied with sea-water, and, with the animals inhabiting the shells, supplies beautiful microscopic objects.

Sertularia operculata (Sea Hair). A common and very elegant species, growing in small dense tufts on shells and seaweed, and, being found in shallow water, more easily obtained with living polypes than some other species.

Sertularia argentea (Squirrel's tail Coralline). Very common, growing on oyster and other shells, and on the roots and stems of the larger fuci. Some of the most beautiful specimens obtained were parasitical on the shells of *Pholas candida*. From its dense mode of growth, and the length and softness of its branches, it has acquired the name of Squirrel's tail Coralline. After high tides or stormy weather it is thrown ashore in large masses along the entire coast. Rarely found with living polypes, except when dredged up from deep water.

Sertularia cupressina (Sea Cypress). Occasionally found on the shore after strong westerly winds; an elegant species, procurable by deep dredging.

Thuiaria Thuia (Bottle-brush Coralline). A very curious species, frequently found on the recess of the tide, growing on shells and stones. It varies much in size, from a few inches to nearly a foot in length, and, except when procured from deep water, is rarely found with living polypes.

Thuiaria articulata (Sea Spleenwort, or Polypody). On shells and stones, sometimes in considerable abundance; at other times not a specimen is to be found on the shore for many weeks together; it is to be obtained from deep water, growing in clusters on the base of *Antennularia*.

We have kept them alive for a considerable time.

Antennularia antennina (Lobster's Horn Coralline, or Sea Beard). Parasitic on shells and stones in deep water, forming dense clusters; it is a beautiful species, and is often dredged up from the scallop bank, off Southport. We have obtained some specimens nearly eighteen inches in length; the colour when first obtained is bright, but it soon changes to a dull horn colour. In the sponge-like base of one of the larger specimens we obtained a number of very minute crabs, some not exceeding hemp-seed in size, and the largest about the third of an inch long, of a brilliant red colour; some of the females had bundles of ova nearly as large as their bodies.

Antennularia ramosa (Branching Lobster's Horn Coralline). This is nearly allied, if not a variety of the preceding. Large specimens are procured from deep dredging in the same localities as the last species.

Plumularia falcata (Sickle Coralline). A beautiful species, abundant on the receding of the tide.

Plumularia cristata (Crested Coralline). Frequently thrown ashore growing on the *Halidrys siliquosa* (the Podded Sea-weed) in great profusion; also on other kinds of fuci, shells, and stones; the vesicles form a beautiful microscopic object, and are readily obtained. They may be gathered in profusion in the months of March and April, and again in the end of August and September; often parasitic on *Plumularia falcata*.

Plumularia setacea. Common on shells and other corallines, and the coarser kinds of fuci, growing in loose tufts. It is found in shallower water than some other kinds, and is consequently more readily obtained with its living inhabitants.

Plumularia pennatula. A very elegant and delicate species, sparingly found on this shore; probably often overlooked from its diminutive size. Fine specimens are found growing on *Cardium aculeatum.*

Plumularia myriophyllum (Pheasant's tail Coralline). This is one of our most beautiful zoophytes when in perfection; the colour is nearly amber, with something of a metallic lustre. It is not common, but we have repeatedly found it at Southport.

Plumularia frutescens (Shrubby Coralline). This species we have several times picked up on the shore; it has a considerable resemblance to small specimens of *Halecium halecinum,* but is darker coloured; the most distinguishing character is its varnished appearance. It attains the height of four inches, but with us seldom exceeds two or three; grows on shells and stones, but is but seldom found, except after severe storms or very heavy spring-tides.

Laomedia dichotoma (Sea-thread Coralline). Growing on stems of fuci, sea-weeds, and on other zoophytes; is of very slender form, but extends a foot or more up the stalks and over the shells on which it is located. Found in large masses, along with its numerous congeners, after strong winds or high tides.

Laomedia geniculata (Knotted Sea-thread Coralline). Much resembling the last species, but generally more upright in growth. Found on the stems of fuci, and, under similar circumstances, with the other species.

Laomedia gelatinosa. Has much resemblance to the other two species, but is generally smaller and more diffuse in its mode of growth; found on shells and the stems of the larger sea-weeds; likewise on sticks that have laid long in the water.

Campanularia volubilis (Small Climbing Coralline). A small but interesting species, frequently growing on other corallines, unoccupied crab and other shells. We have found the shell of the Masked Crab (*Corystes Cassivelaunus*) completely coated with this species.

Campanularia verticillata (Horse-tail Coralline). Occasionally thrown ashore, growing on shells; likewise on the *Tubularia indivisa*.

Alcyonium digitatum (Dead Man's Hands, or Toes; Cow's Paps). Abundant on these shores at most seasons, growing in all kinds of strange and grotesque forms (rarely two specimens alike) on shells, stones, etc. The usual tint is a full cream colour, at times with a faint rosy hue. The general aspect is retained when dried, only somewhat shrunken. To obtain the polypes alive it is necessary to procure it from deep water, where it is found incrusting or attached to stones and shells; we have not been able to keep it alive beyond a few days.

Actinia Mesembryanthemum. From the nature of our shore it would seem a very unlikely locality for any kind of *actinia*, yet several species are occasionally to be found in great plenty, and among them this one is frequently numerous. It is usually found in little pools left by the receding tide, where the sand but slightly covers either clay or peaty soil; we have kept it alive for a considerable time, but it requires a frequent change of water. It is about an inch and a half in diameter, but has the power of depressing itself almost to flatness; the colour varies considerably, from dull pale red to liver colour, streaked and blotched with green and blue—the latter tint predominates. The tentacles are commonly of a paler red than the body, interspersed with some quite blue.

Actinia alba (White Sea Anemone?). Mr. Graves has found specimens which he thinks may be this species, although it is said to be confined to the rocky coasts of Cornwall. The specimens varied from half to three-fourths of an inch in diameter; the colour was dirty white, with white lines or continuous strings of white oblong spots; the tentacles were nearly colourless, with white patches, and in some instances the pellucid tentacles seemed to contain numerous rows of minute white bead-like spots. The *alba* is not so sensitive to the touch as other species, and is found from May to the end of August.

Actinia coriacea. Sometimes found in considerable numbers. It buries itself in the mud and sand, but lives for some time in clear water. The colours are full dull red, blotched and marked with green and dark brown; the tentacles are white, olive, and red intermixed, and when expanded in broad sunshine are truly beautiful, extending considerably beyond the body on all sides, frequently measuring two inches or more in diameter. When at rest the tentacles are all drawn inside, and the exterior surface of the animal so nearly resembles the mud and sand on which it is fixed, that it may readily be overlooked.

Actinia crassicornis. This is the largest of our native Actinias, often attaining four or five inches in diameter.

Actinia Bellis. A beautiful but very diminutive species, variable in appearance and colours. We have been so fortunate as to obtain it once in this locality. The colour is a compound of bright red, white, and ashy grey, with a few yellow spots. Abundant on the Cornish coast, and probably in other places, but, as it is mostly hidden from view by sea-weed, it escapes general observation.

Actinia Dianthus. This beautiful species is very rarely found on this coast; we have only met with two or three, and then in an injured state; they were attached to stones, much bruised, and scarcely evinced any signs of life; they were enabled to retract their external fringe, except partially.

Tubulipora serpens (Small Purple Eschara). Found abundantly on the shore, growing on various zoophytes; in considerable quantities on *Plumularia falcata, Sertularia abietina,* and numerous other species, as well as on shells thrown ashore from deep water.

Crisia eburnea (Tufted Ivory Coralline). On sea-weeds and other zoophytes, forming beautiful little tufts of a clear white colour, frequently in strong contrast with the substance on which it is parasitical.

Crisia denticulata. Larger than the preceding, from which it is principally distinguished by the black joints, though these are not always evident.

Gemellaria loriculata. An abundant species, found at all times on the shore on the receding of the tide; varies from three to eight inches or more in height, growing in dense tufts.

Cellepora pumicosa. Very common, growing on other corallines, stones, and sea-weed; it forms little patches on the stems of *Plumularia falcata, Sertularia abietina,* and various other species, appearing like little pieces of white coral, often slightly tinged with rose colour.

Lepralia. This genus, of which there are from forty to fifty British species, is found on every shore, encrusting shells, stones, zoophytes, and almost every substance thrown ashore by the waves. The forms are very elegant, but being in all cases microscopic, do not receive the general attention the beauty of their appearance deserves.

Cellularia scruposa (Creeping Stony Coralline). Common on shells, corallines, flustra, and the larger sea-weeds. A small species, but interesting for the microscope.

Cellularia reptans (Creeping Coralline). Very common on the same substances and in similar situations to the preceding.

Cellularia plumosa. A large species, often several inches in height, much branched, and frequently found on our shore on the recess of the tide.

Membranipora pilosa. Grows abundantly on other corallines; we have it on *Cellularia scruposa, Plumularia falcata, Sertularia abietina,* and various others.

Membranipora membranacea. Common; often investing *Flustra foliacea.*

Flustra foliacea (Broad-leaved Horn-wrack). This very common zoophyte is noticed by all visitors to the shore, where it is mistaken for a kind of sea-weed. It is of a pale sandy colour, differs considerably in the shape and size of the fronds, and abounds in parasites of various kinds, among which are several other species of *Flustra.*

It inhabits deep water, and is but rarely obtained with living polypes, except when dredged up, attached to shells or stones.

Flustra truncata. Frequently mixed with the above species, which, in general appearance, it much resembles, but is narrower, and the tips of the fronds appear as if cut straight off. Affects the same situations as the common kind.

Flustra avicularis (Bird's-head Horn-wrack). An interesting species, furnished with numerous appendages, closely resembling the head and bill of a bird; and commonly found growing on the *Flustra foliacea.* The aid of a microscope is required to see the construction of the various parts, which well repay a close examination. Not abundant, but may frequently be met with on other flustra, corallines, and shells, from deep water.

Flustra membranacea. Often found coating sea-weed, shells, and other zoophytes. It forms a thin crust which, when dry, readily crumbles to powder.

Salicornaria farciminoides (Bugle Coralline). After tempestuous weather this beautiful species is to be found in considerable abundance, mixed up in the bundles of zoophytes that are left by the retreating tide. It is one of our most elegant corallines, and seems more local than some kinds; it attains two or three inches in height, and often occurs on oyster and scallop shells; is an inhabitant of deep water, and is at times faintly tinted with rose or purple colour.

CHAPTER X.

> There's beauty all around our paths,
>   If but our watchful eyes
> Can trace it 'midst *familiar things*,
>   And through their lowly guise.
>
> HEMANS.

FORAMINIFERA OF SOUTHPORT.

THE Foraminifera are members of that class of animal life termed *Rhizopoda*, or root-footed animals, so named from the power they possess of putting out filaments for the purpose of locomotion; or, as it were, extemporising limbs from any part of their body as occasion requires. Collectively, the *Rhizopoda* form one of the three classes into which the *Protozoa*, the lowest forms of life, are divided, the other classes being the *Porifera*, the type of which is the common sponge, and the *Infusoria*. There are two distinct orders in the *Rhizopoda*, the fresh-water and the marine. The first is well represented by the *Amœba*, common in fresh-water ponds, and in describing which we describe the animals of the whole class. The *Amœba* is a minute jelly-like substance, without any differentiation of parts; a simple homogeneous mass, without apparent organisation, capable of changing into a great variety of forms, laying hold of its food without members, swallowing it without a mouth, digesting it without a stomach, appropriating its nutritious material without absorbent vessels, or a circulating system, moving from place to place without muscles, feeling, if it has the power to do so, without nerves, but in many instances forming shells of a symmetry and complexity not surpassed by those of any testaceous animals. Although the *Amœba* has no stomach, digestive cavity, or mouth, it is exceedingly voracious. Throwing out its filaments or *pseudopodia*, it gradually, with their aid, crawls along the bottom of the pond, until it comes in contact with any object suitable for food, such as a diatom or a particle of vegetable matter, when the animal at once commences to envelope the object with its own substance, whereby it becomes not exactly swallowed, but embedded. When all the digestible matter has been absorbed, the residue is ejected, indifferently from any part, wherever it happens to be nearest to the exterior.

Several genera of the fresh-water *Rhizopoda* form a horny case, in which there are openings for the protrusion of their *pseudopodia*. The genus *Difflugia* forms a transparent horny case, and fixes upon the outside, particles of sand, and the siliceous shells of *Diatomaceæ*. Although the animal is of such excedingly low organisation, it exhibits a curious power of choice, never by any chance selecting other than mineral particles; fixing them, probably when the case is in a certain glutinous condition, with an appearance of order; in some displaying almost the symmetry of design, smaller grains being selected and neatly fitted into the interstices of larger particles. The sarcode is of such non-consistency that the filaments in coming in contact with one another coalesce when there are several points of contact, forming a rough network, all being retracted into the substance of the animal at its will.

The Foraminifera themselves are wholly marine. The physiological characteristics of the animals are apparently the same as in the fresh-water orders. Food is embedded and digested, and progression is effected by the aid of pseudopodia. The generic differences consist in the animal forming shells of many beautiful, curious, and varied forms. They are found in abundance in all seas, the maximum of development being in the torrid, the minimum in the frigid zone. They are commonly found alive upon sea-weeds; every dredging from the bottom will bring them up. I have found Foraminifera in great abundance upon the Southport shore amongst the fine débris left at every tide mark. They are brought up, almost unmixed with sand or mud, when sounding the lowest known depths of the Atlantic, where, until very recently, it was supposed animal life could not exist, on account of the pressure of the mass of water. It is, however, a fact that they are found in the greatest abundance at the lowest depths, but they are then fewer in species or variety than in shallower waters. Ehrenberg says that chalk is composed, in a great proportion, of their shells, and that they are found to be the principal, or a very large constituent of whole mountain ranges of rocks. We thus see that one of the lowest forms of organic life of the present day has been continued from vastly remote geological ages, whilst in the higher and more developed forms of life we have only analogical and modified resemblances.

In classifying the Foraminifera the first great distinction which strikes the observer is their division into one-celled (*Monothalamia*) and many-celled (*Polythalamia*). The animals, being all of one sarcode-like substance, exhibiting the same habits of life, cannot—as with the higher orders, such as the Mollusca—be taken into any, or for more than a very little account for generic or specific distinction. The characters of the shell are therefore looked to as a basis of classification, the form and arrangement of the cells or chambers, the mode of their connection, the presence or absence of large or small openings, their markings, and the intimate structure of the shell, are the principal points for consideration. The Foraminifera are thus arranged into families, genera, species, and varieties. Our knowledge of them, it must be confessed, is unsatisfactory and imperfect; gradually, however, no doubt they will be better understood, as they are now receiving a considerable amount of attention from eminent naturalists. Theories and classifications are, by observation and research, being continually modified and altered; but one fact is universally observed, and that is the variableness of these minute creatures. In several species a definite type of form prevails, or rather the idea of it, whilst scarcely two shells will be found identically the same. I have noticed this particularly in the *Miliolinæ*, which are exceedingly abundant on our shore. Dr. Carpenter says, "The range of variation is so great among Foraminifera as to include not merely the differential characters which systematists, proceeding upon the ordinary methods, have accounted specific, but also those upon which the greater part of the genera of this group have been founded, and even in some instances those of its orders. The ordinary notion of species, as assemblages of individuals, marked out from each other by definite characters, that have been generically transmitted from original prototypes, similarly distinguished, is quite inapplicable to this group, since, even if the limits of such assemblages were extended so as to include what would elsewhere be accounted genera, they would still be found so intimately connected by gradational links that definite lines could not be drawn between them." He also propounds the idea that the many generic and specific forms may be traced up to a very few original and leading types, which have multiplied by diversities of temperature, depth, geological position, and local influences, thus exhibiting in this group, to a certain extent, Mr. Darwin's theory of multiplication of species by natural selection. This theory receives a certain amount of confirmation when we notice the comparatively few species found in the profound depths of the ocean, where the disturbing influences are small; and the great number of apparent species and varieties found in shallower waters, where disturbing influences of every kind are constantly in operation.

The propagation and development of the *Rhizopods* are not yet definitely defined. Further observations of the living animal are required before we can read their life-history with certainty; the information we possess leads to the inference that they are propagated by fission. The *Amœba* has been observed to throw out a pseudopod, a little thickened at the end, and attach it to an object; then, without drawing the body forward, the filament has gradually become thinner until the enlarged point has become detached, and like the original body, has thrown out pseudopodia of its own. Similar phenomena have been observed in the Foraminifera, or the shell-bearing class. Although the whole process of growth and development cannot be traced, there is reason to believe that the animal arrives at a mature size before commencing to form its shell. Unlike the Mollusk, which commences its free life with a shell, and progressively increases its dimensions by the deposition of carbonate of lime, the Foraminifera form a transparent shell of mature size, which it gradually thickens, when we read their comparative ages by contrasting their opacity.

Southport is a very favourable locality for the collection of Foraminifera; in a comparatively short time I have been enabled to add considerably to the number of species recorded by Professor Williamson as occurring on this coast.

For the benefit of the would-be collector I may state the best method of collecting. In walking along the shore, at either high or low-water mark, little slopes are seen, where the retreating wave has left fine débris of comminuted shells, etc. Carefully scrape this off the sand, take home and thoroughly dry it, then throw a little at a time on a basin of water, when the heavy particles will sink and leave the Foraminifera floating on the top. Skim this off, dry it, and examine under a lens. The objects sought for will then be seen, and may be picked out with the point of a needle. Touch your hair with the needle—this slightly greases it and causes the Foraminifera to adhere lightly to the point. Fix the needle into a little wooden handle for convenience in using.

LIST OF THE SOUTHPORT FORAMINIFERA.

Orbulina universa.
Lagena vulgaris.
Lagena vulgaris var. *clavata.*

Lagena vulgaris var. *pellucida.*
Lagena vulgaris var. *semistriata.*
Lagena vulgaris var. *striata.*
Lagena vulgaris var. *interrupta.*
Lagena vulgaris var. *gracilis.*
Lagena vulgaris var. *substriata.*
Entosolenia squamosa.
Nodosaria radicula.
Nodosaria pyrula.
Dentalina subarcuata.
Dentalina legumen.
Cristellaria subarcuatula.
Nonionina Jeffreysii.
Nonionina elegans.
Polystomella umbilicatula.
Polystomella umbilicatula var. *incerta.*
Polystomella crispa.
Rotalina Beccarii.
Rotalina inflata.
Rotalina concamerata.
Rotalina nitida.
Rotalina fusca.
Planorbulina vulgaris.
Truncatella lobata.
Bulimina pupoides.
Bulimina pupoides var. *fusiformis.*
Cassidulina lævigata.
Polymorphina lactea.
Polymorphina lactea var. *oblonga.*
Polymorphina lactea var. *flatulosa.*

Polymorphina lactea var. *communis.*
Biloculina ringens.
Biloculina ringens var. *carinata.*
Biloculina ringens var. *patagonica.*
Spiroloculina depressa.
Miliolina seminulum.
Miliolina seminulum var. *oblonga.*
Miliolina seminulum var. *disciforn*
Miliolina bicornis.
Miliolina bicornis var. *angulata.*
Miliolina trigonula.
Spirilina foliacea.

MEAN MONTHLY AND ANNUAL TEMPERATURE IN THE SHADE.

	1872	1873	1874	1875	1876	1877	1878	1879	1880	1881	1882	Mean
	Degrees	Degrees	Degrees	Degrees	Degrees	Degrees	Degrees	Degrees	Degrees	Degrees	Degrees	Degrees
January	41.0	42.0	42.6	42.6	38.3	41.2	40.2	30.8	34.2	28.6	41.9	38.5
February	43.7	36.1	38.7	36.5	39.9	43.1	40.9	36.4	40.8	36.5	43.0	39.6
March	44.2	41.4	43.7	39.6	40.1	40.1	41.7	39.0	41.1	39.7	45.3	41.4
April	46.5	46.4	49.3	47.1	45.4	41.0	48.2	41.5	45.0	42.6	46.5	45.4
May	49.3	49.2	49.1	52.3	48.3	47.3	52.6	47.2	48.6	49.9	51.9	49.6
June	56.2	57.8	56.1	56.2	56.2	57.8	57.5	54.9	54.8	53.2	55.0	56.0
July	61.7	60.6	61.6	58.3	60.3	57.8	60.9	56.1	57.7	57.3	58.3	59.1
August	60.0	59.7	58.8	60.2	59.7	58.3	60.3	56.7	59.6	54.9	58.1	58.8
September	56.6	53.9	55.5	58.1	54.3	52.1	56.0	53.3	56.7	52.4	53.8	54.8
October	47.6	48.4	50.5	48.1	51.6	49.1	50.1	47.9	43.7	45.4	50.2	48.4
November	44.5	43.0	42.5	41.7	42.1	45.3	38.2	39.9	40.5	47.4	42.8	42.5
December	41.6	43.0	32.1	39.9	41.8	41.6	31.7	32.5	39.9	39.6	38.4	38.4
Mean	49.4	48.5	48.4	48.4	48.2	47.9	48.2	44.7	46.9	45.6	48.8	47.7

MEAN MONTHLY AND ANNUAL DAILY RANGE OF TEMPERATURE.

	1872	1873	1874	1875	1876	1877	1878	1879	1880	1881	1882	Mean
	Degrees	Degrees	Degrees	Degrees	Degrees	Degrees	Degrees	Degrees	Degrees	Degrees	Degrees	Degrees
January	8.2	8.5	8.2	8.4	9.9	9.7	7.6	9.9	11.4	14.1	7.3	9.4
February	9.0	8.0	10.1	7.8	9.5	7.7	9.9	7.9	10.2	10.7	7.7	8.9
March	11.5	12.3	10.4	9.7	10.5	10.8	9.6	10.9	17.5	12.7	9.5	11.4
April	11.2	13.4	14.0	14.9	13.3	11.4	14.4	12.5	13.9	15.9	12.2	13.4
May	10.0	11.8	13.1	12.7	14.9	13.6	12.0	11.8	15.9	18.9	14.6	13.6
June	11.3	11.4	14.8	12.3	13.2	14.7	13.5	12.0	14.2	15.9	10.2	13.1
July	12.4	12.9	13.3	11.2	12.4	9.5	12.7	9.7	11.9	13.3	11.1	11.8
August	11.7	9.4	11.3	12.0	13.4	11.2	12.3	11.9	16.5	12.5	10.1	12.0
September	8.6	11.6	12.5	12.6	10.7	12.3	11.4	13.1	14.2	15.0	11.0	12.1
October	10.5	10.9	9.9	9.3	11.0	12.2	11.1	10.9	14.6	12.9	9.3	11.1
November	8.5	8.6	9.0	8.3	9.4	9.8	9.9	11.2	14.6	9.3	8.6	9.8
December	7.0	8.5	9.1	7.2	8.7	8.5	12.2	13.9	12.5	8.7	6.7	9.4
Mean	10.0	10.6	11.3	10.5	11.4	10.9	11.4	11.3	13.9	13.3	9.9	11.3

MONTHLY RANGE OF TEMPERATURE IN THE SHADE.

	1872	1873	1874	1875	1876	1877	1878	1879	1880	1881	1882	Mean
	Degrees	Degrees	Degrees	Degrees	Degrees	Degrees	Degrees	Degrees	Degrees	Degrees	Degrees	Degrees
January	26.0	27.3	23.0	38.1	30.7	26.7	29.3	31.0	36.3	44.2	19.9	30.2
February	28.2	30.4	29.5	27.1	33.5	25.2	34.1	24.3	29.7	26.6	24.1	28.4
March	37.9	34.3	37.6	27.7	30.7	31.6	26.7	26.3	30.8	31.7	24.0	30.8
April	30.6	40.8	40.9	38.7	37.4	24.0	38.8	30.8	31.6	40.1	28.0	34.7
May	28.9	27.5	33.0	31.7	30.4	37.6	27.7	31.0	34.1	40.5	36.9	32.7
June	40.1	29.1	30.1	33.8	40.0	36.0	43.8	29.7	36.9	33.0	24.5	34.3
July	33.3	43.4	33.8	29.4	32.4	19.6	38.1	25.0	23.6	28.9	22.6	30.0
August	30.7	24.7	24.1	27.3	46.9	25.6	28.4	29.5	32.1	37.8	29.0	30.5
September	29.3	36.9	38.1	32.4	35.1	23.8	34.5	36.0	44.4	27.5	25.0	33.0
October	29.6	37.1	22.7	30.8	32.4	30.6	36.4	32.1	41.2	32.0	35.0	32.7
November	30.8	26.4	30.8	29.2	33.1	25.9	23.6	33.7	37.3	26.2	31.2	29.8
December	27.1	27.7	36.9	28.8	27.4	21.9	40.7	43.2	29.1	27.1	32.0	31.1
Mean	31.1	32.1	31.7	31.2	34.2	27.4	33.5	31.0	33.9	33.0	27.7	31.5

MEAN DEGREE OF HUMIDITY OF THE AIR AT 9 A.M.

	1872	1873	1874	1875	1876	1877	1878	1879	1880	1881	1882	Mean
	Degrees	Degrees	Degrees	Degrees	Degrees	Degrees	Degrees	Degrees	Degrees	Degrees	Degrees	Degrees
January	89	85	86	90	89	90	89	86	85	81	88	87
February	92	86	88	85	87	85	90	89	88	89	89	88
March	85	84	85	82	83	86	83	84	85	86	87	84
April	81	74	73	72	83	82	82	82	82	81	84	80
May	76	77	73	79	76	74	82	78	78	77	82	77
June	80	76	69	76	75	73	84	77	85	80	81	78
July	78	76	71	78	83	80	81	84	88	81	86	81
August	80	79	81	80	82	82	86	86	84	84	83	82
September	80	80	81	82	84	81	87	84	87	84	85	83
October	86	84	83	84	87	84	86	90	84	81	88	85
November	86	86	84	85	91	86	84	87	86	89	87	87
December	88	88	86	87	93	91	90	85	87	83	89	88
Mean	83	81	80	82	84	83	85	84	85	83	86	83

MONTHLY AND ANNUAL AMOUNTS OF RAINFALL.

	1872	1873	1874	1875	1876	1877	1878	1879	1880	1881	1882	Mean
	Inches	Inches	Inches	Inches	Inches	Inches	Inches	Inches	Inches	Inches	Inches	Inches
January.	4.529	3.820	2.664	4.610	1.339	6.288	3.034	1.546	0.580	0.655	2.065	2.830
February	2.802	0.499	1.408	0.951	3.083	2.910	1.497	3.046	2.014	4.560	1.499	2.206
March...	3.584	3.221	2.138	0.852	2.084	2.370	1.542	1.590	2.757	3.363	2.674	2.380
April ...	1.876	0.512	1.154	1.150	2.337	2.292	1.787	1.965	1.105	1.781	4.991	1.905
May	2.388	1.743	1.688	1.928	0.664	2.454	3.500	2.234	2.256	3.717	1.787	2.214
June.....	4.281	1.702	1.332	2.951	2.836	1.425	5.069	3.225	2.753	3.002	4.237	2.983
July......	7.374	3.473	1.867	5.260	2.907	4.362	0.536	4.195	2.847	3.882	5.133	3.803
August..	2.816	2.711	3.735	3.781	5.060	6.086	5.460	6.673	1.793	5.698	4.061	4.352
Septem'r	6.562	2.279	3.038	6.484	3.477	4.990	3.655	3.032	3.140	2.193	3.067	3.811
October.	5.141	4.325	5.774	7.060	3.178	4.565	3.855	1.742	2.966	3.448	3.521	4.143
Novem'r.	3.342	1.802	4.494	5.135	3.828	5.387	2.450	0.953	3.359	3.774	4.916	3.585
Decem'r.	3.783	1.182	2.380	1.574	4.567	3.848	1.782	2.130	5.894	4.548	2.390	3.098
Total..	48.478	27.269	31.672	41.736	35.360	46.977	34.167	32.331	31.464	40.621	40.341	37.310

MEAN ANNUAL AMOUNTS OF OZONE.

1872.........5.60	1876.........4.51	1880.........4.72
1873.........5.40	1877.........6.01	1881.........4.85
1874.........4.91	1878.........5.30	1882.........4.94
1875.........4.21	1879.........4.96	

MEAN MONTHLY AMOUNTS OF OZONE

From 9 p.m. to 9 a.m., or night period; and from 9 a.m. to 9 p.m., or day period; and the Means of the two periods for the years 1872–82.

	Night Mean.	Day Mean.	Mean of Day and Night.
January	4.16	3.43	3.79
February	4.93	4.17	4.55
March	5.66	5.45	5.55
April	5.17	5.61	5.39
May	5.95	6.61	6.28
June	5.75	6.75	6.25
July	5.65	6.68	6.16
August	5.17	6.01	5.59
September	4.72	5.09	4.90
October	3.86	3.60	3.73
November	4.80	4.11	4.45
December	4.15	3.47	3.81
Means	5.00	5.08	5.04

POPULATION OF SOUTHPORT.

Year.		Population.
1861	Census	8,940
1862	Estimated	9,683
1863	,,	10,293
1864	,,	11,054
1865	,,	11,851
1866	,,	12,716
1867	,,	13,644
1868	,,	14,640
1869	,,	15,709
1870	,,	16,856
1871	Census	18,086
1872	Estimated	18,846
1873	,,	19,638
1874	,,	20,463
1875	,,	23,113
1876	,,	25,997
1877	,,	27,132
1878	,,	28,317
1879	,,	29,554
1880	,,	30,845
1881	Census	32,191
1882	Estimated	33,597
1883	,,	35,065

Aggregate Population of Southport and Birkdale (Census 1881) 42,454.

COMPARATIVE DEATH-RATE TABLE.

The death-rate in England and Wales from all causes was,
in 1881..18·90
in 1882......... ....19·60

The death-rate in Southport from all causes was,
in 1881..17·64
in 1882..16·43

The death-rate in England and Wales from the seven principal zymotic diseases was, in 1881 ....2 24
in 1882 ....2·64

The death-rate in Southport from the seven principal zymotic diseases was, in 1881 ....1·36
in 1882 ....1.01

The proportion of deaths under one year of age to 1,000 births in England and Wales was, in 1881...................130
in 1882...................141

The proportion of deaths under one year of age to 1,000 births in Southport was, in 1881 100
in 1882 119

INDEX.

	PAGE
AQUARIUM, THE	10
ARACHNIDA OF SOUTHPORT	103
ART GALLERY, THE	8
ATKINSON, LATE MR. WILLIAM	8
ATMOSPHERE OF SOUTHPORT	31, 40
BANKS	15
BARTON, THE LATE MR.	4
BATHS, THE VICTORIA	17
BAXENDELL, MR. WILLIAM	27, 31
BIRDS, THE SOUTHPORT	89
BIRKDALE	5
BLYTHE HALL	2
BOTANICAL GARDENS, CHURCHTOWN	11
BRONCHITIS	39
CAMBRIDGE HALL, THE	8
———— LOCAL EXAMINATIONS	19
CATHOLIC CHAPELS	14
CEMETERY, THE	15
CHRIST CHURCH	5
CHURCHES, SOUTHPORT	13
CHURCHTOWN	2, 5, 14
———— BOTANICAL GARDENS	11
CLIMATE OF SOUTHPORT	27, 39
COCKLES	65
CONCERT HALL, THE	10, 11
CONGREGATIONALISTS, THE	14
CONSERVATORY, THE CHURCHTOWN	12
———— WINTER GARDENS	10
CONSUMPTION	36
CONVALESCENT HOME, THE MANCHESTER AND SALFORD	17
———— HOME FOR CHILDREN	18
———— HOSPITAL, THE	16
CROSSENS	81, 86, 93
CRUSTACEA, SOUTHPORT	111
DEATH-RATE OF SOUTHPORT	173
DRAINAGE OF SOUTHPORT	20

INDEX.

	PAGE
Duke's Folly	4
Dyspepsia	42
Fernley, the late Mr. John	9
Ferns, Southport	86
Fishermen, Southport	34
Fleetwood Family, the	14
Flora, the Southport	65
Foraminifera, the Southport	158
Franchise, the Southport	19
Free Library	8, 9
Geology of Southport District	21
Gentlewomen's Home, the	18
Glaciarium, the	12
Government of Southport	18
Governesses' Home, the	18
Graves, the late Mr.	94
Gulls, the Sea	93, 100
Hawes, the	3
Hesketh Family, the	14
——— Rev. Charles	11, 14
——— Park, the	11
Holden, Mr. Edward	13
Hotels, Southport	20
Hydropathic Establishments	20
Incorporation of Southport	18
Infirmary, the	17
Invalids, Suggestions for	35
Lathom House	2
Literary and Philosophical Society	9
Lord-street	5
Market, the	15
Meteorology of Southport	27
——— Tables	166
Mollusca, Southport	114
Museum, the Botanic Gardens	12, 93
Natural History of Southport	64
Newspapers, Southport	18
North Meols	3
——— Rector of	14
Ornithology	89
Ozone	30

INDEX.

	PAGE
Ozone Tables	171
Pickard-Cambridge, Rev. O.	103
Pier, the	7
Population of Southport	5, 172
Promenade, the	6
Railways, Southport	2, 18
Rainfall, the Southport	27, 32, 168
Rent of Houses	19
Rufford Hall	2
Sanatorium, Southport as a	35
Sandhills, the	22, 67
Sands, the	8, 29
Scarisbrick Hall	2
Schools	19
Sea-Bathing	54
———— Infirmary	16
Shells found at Southport	118
Strangers' Charity, the	16
St. Cuthbert's Church	13
Sunsets, Southport	7
Sutton, the late Mr. William	3
Temperature of Southport	27, 43, 166
Town Hall, the	8
Tyrer, the late Mr.	94
Waterworks, the	19
Wesleyan Chapels	14
Winds, Prevailing	28
Winter Gardens, the	9
Zoophytes, the Southport	140

www.ingramcontent.com/pod-product-compliance
Lightning Source LLC
Chambersburg PA
CBHW032151160426
43197CB00008B/859